Advanced Circuit Simulation
Using Multisim Workbench

Synthesis Lectures on Digital Circuits and Systems

Editor
Mitchell A. Thornton, *Southern Methodist University*

The Synthesis Lectures on Digital Circuits and Systems series is comprised of 50- to 100-page books targeted for audience members with a wide-ranging background. The Lectures include topics that are of interest to students, professionals, and researchers in the area of design and analysis of digital circuits and systems. Each Lecture is self-contained and focuses on the background information required to understand the subject matter and practical case studies that illustrate applications. The format of a Lecture is structured such that each will be devoted to a specific topic in digital circuits and systems rather than a larger overview of several topics such as that found in a comprehensive handbook. The Lectures cover both well-established areas as well as newly developed or emerging material in digital circuits and systems design and analysis.

Advanced Circuit Simulation Using Multisim Workbench
David Báez-López, Félix E. Guerrero-Castro, and Ofelia Cervantes-Villagómez
2012

Circuit Analysis with Multisim
David Báez-López and Félix E. Guerrero-Castro
2011

Microcontroller Programming and Interfacing Texas Instruments MSP430, Part I

Steven F. Barrett and Daniel J. Pack
2011

Microcontroller Programming and Interfacing Texas Instruments MSP430, Part II
Steven F. Barrett and Daniel J. Pack
2011

Pragmatic Electrical Engineering: Systems and Instruments
William Eccles
2011

Pragmatic Electrical Engineering: Fundamentals
William Eccles
2011

Introduction to Embedded Systems: Using ANSI C and the Arduino Development Environment
David J. Russell
2010

Arduino Microcontroller: Processing for Everyone! Part II
Steven F. Barrett
2010

Arduino Microcontroller Processing for Everyone! Part I
Steven F. Barrett
2010

Digital System Verification: A Combined Formal Methods and Simulation Framework
Lun Li and Mitchell A. Thornton
2010

Progress in Applications of Boolean Functions
Tsutomu Sasao and Jon T. Butler

2009

Embedded Systems Design with the Atmel AVR Microcontroller: Part II
Steven F. Barrett
2009

Embedded Systems Design with the Atmel AVR Microcontroller: Part I
Steven F. Barrett
2009

Embedded Systems Interfacing for Engineers using the Freescale HCS08
Microcontroller II: Digital and Analog Hardware Interfacing
Douglas H. Summerville
2009

Designing Asynchronous Circuits using NULL Convention Logic (NCL)
Scott C. Smith and JiaDi
2009

Embedded Systems Interfacing for Engineers using the Freescale HCS08
Microcontroller I: Assembly Language Programming
Douglas H. Summerville
2009

Developing Embedded Software using DaVinci & OMAP Technology
B.I. (Raj) Pawate
2009

Mismatch and Noise in Modern IC Processes
Andrew Marshall
2009

Asynchronous Sequential Machine Design and Analysis: A Comprehensive
Development of the Design and Analysis of Clock-Independent State
Machines and Systems
Richard F. Tinder

2007

PSpice for Digital Communications Engineering
Paul Tobin
2007

PSpice for Circuit Theory and Electronic Devices
Paul Tobin
2007

Pragmatic Circuits: DC and Time Domain
William J. Eccles
2006

Pragmatic Circuits: Frequency Domain
William J. Eccles
2006

Pragmatic Circuits: Signals and Filters
William J. Eccles
2006

High-Speed Digital System Design
Justin Davis
2006

Introduction to Logic Synthesis using Verilog HDL
Robert B. Reese and Mitchell A. Thornton
2006

Microcontrollers Fundamentals for Engineers and Scientists
Steven F. Barrett and Daniel J. Pack
2006

Advanced Circuit Simulation Using Multisim Workbench

David Báez-López, Félix E. Guerrero-Castro, and Ofelia Cervantes-Villagómez

ISBN: 978-3-031-79842-9   paperback
ISBN: 978-3-031-79843-6   ebook

DOI 10.1007/978-3-031-79843-6

A Publication in the Springer series
SYNTHESIS LECTURES ON DIGITAL CIRCUITS AND SYSTEMS

Lecture #36
Series Editor: Mitchell A. Thornton, *Southern Methodist University*
Series ISSN
Synthesis Lectures on Digital Circuits and Systems
Print 1932-3166   Electronic 1932-3174

Advanced Circuit Simulation Using Multisim Workbench

David Báez-López, Félix E. Guerrero-Castro,
and Ofelia Cervantes-Villagómez
Universidad de las Américas–Puebla

SYNTHESIS LECTURES ON DIGITAL CIRCUITS AND SYSTEMS
#36

ABSTRACT

Multisim is now the de facto standard for circuit simulation. It is a SPICE-based circuit simulator which combines analog, discrete-time, and mixed-mode circuits. In addition, it is the only simulator which incorporates microcontroller simulation in the same environment. It also includes a tool for printed circuit board design.

Advanced Circuit Simulation Using Multisim Workbench is a companion book to Circuit Analysis Using Multisim, published by Morgan & Claypool in 2011. This new book covers advanced analyses and the creation of models and subcircuits. It also includes coverage of transmission lines, the special elements which are used to connect components in PCBs and integrated circuits. Finally, it includes a description of Ultiboard, the tool for PCB creation from a circuit description in Multisim. Both books completely cover most of the important features available for a successful circuit simulation with Multisim.

KEYWORDS

circuit simulation, electrical circuits, electronic circuits, subcircuits, models, transmission lines, temperature analysis, sensitivity analysis, noise, Monte Carlo analysis, worst-case analysis, microcontroller simulations, printed- circuit board design

Contents

Preface

This book is a companion for the Circuit Analysis and Simulation book published in 2011. That book covers basic analyses with Multisim. This book is concerned with advanced analyses. That is, those analyses that are more specialized and that can be done after a circuit has been tested with one or more of the types of analyses covered in the companion book.

The book starts by presenting a way to modify models and offering two techniques to create subcircuits. Thus, users can build their own libraries containing parts, models, and subcircuits. The second chapter covers transmission lines. They are very important in the simulation of integrated circuits, but they can be used to simulate other kind of circuits like, for example, digital filters. The third chapter covers advanced topics such as sensitivity, noise, Monte Carlo and worst-case, parametric, and temperature analyses. Each of these analyses is usually carried out after any one of the basic analyses has been finished and the behavior of the circuit is satisfactory to the designer. Thus, many of the characteristics of the circuit can be thoroughly evaluated. The fourth chapter is devoted to the simulation of microcontrollers. This is a unique feature to Multisim, that is, the incorporation of microcontrollers in a circuit simulation together with analog components in a mixed-mode fashion. Two of the most popular microcontrollers are available in the microcontroller libraries. The last chapter is devoted to printed circuit board (PCB) design using Ultiboard, which is a tool especially designed for this purpose. Two examples illustrate the procedure for the PCB design.

Two of the authors (DBL and OCV) acknowledge the help provided by our undergraduate students Carmen María Saavedra, Gerardo de la Rosa, Miguel Angel Juárez Rivas, Ricardo Gómez Crespo, and Daniel Treviño García for simulating each one of the circuits and providing valuable feedback to improve the book.

The authors also thank their families for the continued support during the writing of both books, for their patience and understanding.

David Báez-López, Félix E. Guerrero-Castro, and Ofelia Cervantes-Villagómez
February 2012

CHAPTER 1

Models and Subcircuits

Resistors, capacitors, inductors, diodes, and transistors are described in Multisim by a model. Other devices are rather described by a subcircuit.

A model is a description of the device by using its defining equations. Thus, a model can be built for any device whose equations are available from a theoretical analysis of its behavior and construction. Diodes and transistors are examples of circuit elements described by a model.

A subcircuit is a smaller circuit which can represent a set of specific properties of a larger circuit. Operational amplifiers and digital circuits such as flip-flops and gates are examples of circuits defined by subcircuits. A subcircuit is the equivalent of a method in object oriented programming and, thus, it can be reused whenever we require it.

Multisim libraries contain a great deal of parts defined by subcircuits. The definition of a subcircuit in the libraries is transparent to the user and can only be appreciated if we open the model for a device.

A user can modify a device model, can create a subcircuit to be used in a larger circuit, and can also create a subcircuit and make it available at either the Corporate or the User Database to be used by any circuit designed later on. In this section we present the procedures for the following:

1. Modify an existing element's model.

2. Create a subcircuit within a larger circuit, and

3. Create a subcircuit and make it available at the database.

We show with three examples the procedures to accomplish these three tasks.

1.1 EDITING A COMPONENT MODEL IN MULTISIM

Multisim has the capability of editing components available in the Master Database. This is useful when we need to fit components to specific needs. For example, change the W/L ratio in an MOS transistor, the β in a BJT, the input impedance of an op amp, etc. There are two ways to edit components. The first method edits the component in such a way that they can be only used within the circuit where it was edited. It is not available for any other circuit simulation. The second method edits a component and then places it in the database for use in any other circuit simulation.

1.1.1 EDITING A COMPONENT FOR USE IN THE SAME CIRCUIT ONLY

We show this procedure with an example. In the example we edit the β of the transistor. Since the emitter resistor is missing, the amplifier is very sensitive to changes in β.

Example 1.1   Editing of BF in a common emitter amplifier.
Let us consider the common emitter amplifier shown in Figure 1.1. This circuit uses a 2N3904 bipolar transistor.

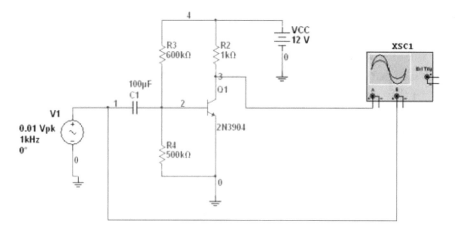

Figure 1.1: Common emitter amplifier.

From the oscilloscope window in Figure 1.2 we can see that the amplifier gain is given by:

$$A_v = \frac{V_{out}}{V_{in}} = \frac{-550.090\text{mV}}{5.735\text{mV}} = -95.91$$

Now, we edit the model for the 2N3904 bipolar transistor.

To edit this component we double click on the bipolar transistor 2N3904 to open the window of Figure 1.3. In this window we click on the Edit Model button to obtain the Edit Model window of Figure 1.4. In this window we see that the Bf = 416.4. We change this value to 50.0 and press enter. The new value is shown in Figure 1.5.

We now click on the Change Part Model to return to the window of Figure 1.3 and press the OK button. The circuit now looks as shown in Figure 1.6. We note an asterisk following the transistor number. This means that its model has been modified. We run the analysis again by pressing the Run button and take a look at the output shown in the oscilloscope as shown in Figure 1.7. We measure the voltage gain and it has changed to

$$A_v = \frac{206.878\text{ mV}}{-7.103\text{ mV}} = -29.13$$

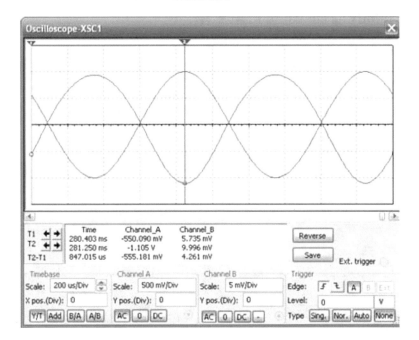

Figure 1.2: Input and output waveforms for the common emitter amplifier.

which is less than the value obtained in the case of $\beta = 416.4$ where we used the original value for β.

1.1.2 EDITING A COMPONENT IN THE DATABASE

Multisim has all the components grouped in databases. The available databases are the following: the Master database, the Corporate database, and the User database. We can only edit components in the Corporate and the User databases. Thus, if we wish to edit a model in the Master database, we have first to copy the component to any of the other two databases.

Example 1.2 Copying a component from the Master Database to either the Corporate or User database.
In this example we want to copy a component from the Master database to the User database. We use the transistor 2N3904 from Example 1.1. In this circuit, or in any circuit with the transistor, we select it and then we select Tools→Database→Save Component to Database. This opens the dialog window shown in Figure 1.8. We select the User Database and the transistors family as shown in Figure 1.9.

Figure 1.3: Dialog window for the bipolar transistor.

Then we click on the Add family button and the dialog window shown in Figure 1.10 opens. There we change the default name Def to Bipolar, as is shown in the figure. We press the OK button. This closes the window and it returns to the Select Destination Family window, and the new family group is displayed, as shown in Figure 1.11. When we click on the OK button, a message indicates the successful copying of the element to the target database and the window is closed, and the Database Manager window shown in Figure 1.12 is open, showing the User Database with the transistor 2N3904 as the only element available in the database. We are now ready to edit this transistor and change its parameters.

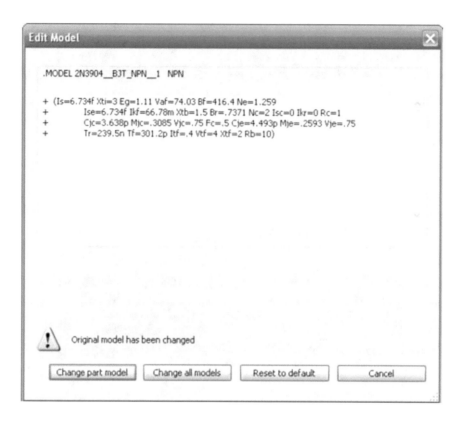

Figure 1.4: Edit model window for the 2N3904.

Example 1.3   Editing the 2N3904 bipolar transistor model in the User database.

We continue with Example 1.2. Here we are going to modify the β of the transistor and then we test the changes in the common emitter amplifier from Example 1.1. In order to do so we have to modify the model of the bipolar transistor 2N3904. To achieve our goal we have to complete the following steps.

1. To edit the model we select Tools→Database→Database Manager to obtain the window in Figure 1.14. This figure is identical to Figure 1.12 because is the same component we have not added new components. If there are more components in the database, we look for the required component. Once this is done we press the Edit button. This opens the Component Properties window of Figure 1.15. There

we see that the value for the parameter Bf is 416. To edit the model we press on the Add/Edit button which opens the dialog window of Figure 1.16.

Figure 1.5: Change of the parameter Bf to Bf = 50 for the 2N3904.

We proceed to make the desired changes. In this example we wish to change Bf from its nominal value of 416.4 to the desired value of 85, as shown in Figure 1.17. We press the Select button in the same window and Multisim warns us that the model has changed and if we wish to proceed, as shown in Figure 1.18 and then, after clicking the OK, Multisim takes us to the window of Figure 1.19 where we press the Overwrite button. This opens the Select a Model window of Figure 1.20 which shows the model for our component.

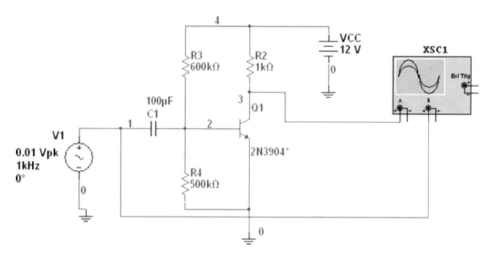

Figure 1.6: Transistor with Bf = 50.

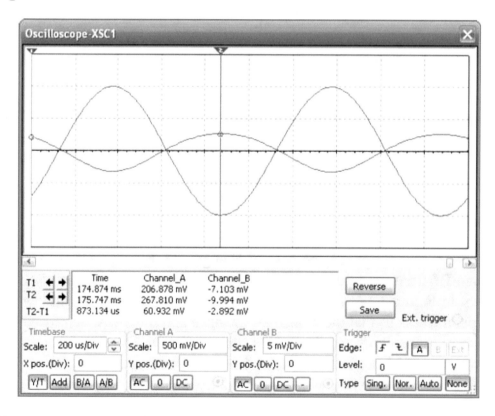

Figure 1.7: Input and output waveforms for amplifier with BF=50. The gain has decreased because BF has decreased.

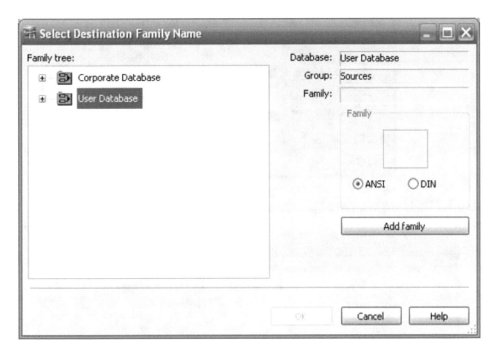

Figure 1.8: Dialog window to select the destination database.

Figure 1.9: Expansion of the User database with the Transistors family selected.

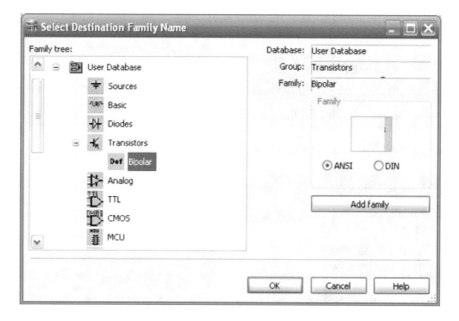

Figure 1.10: Here we name the new family group.

Figure 1.11: Expansion of the User database with the new family displayed.

If we wish to rename the component, we can do it in the Component Properties window. This window can be open by double clicking on the transistor. This opens the window for the component (see Figure 1.3) and we click on the Edit Component in DB which opens the Component Properties window. There we can change the name as desired. We wish to change the model's name to 2N3904_new, as shown in Figure 1.21.

This takes us to the Select Destination Family Name window as it was done before. When we are finished, we are ready to simulate circuits with this new component.

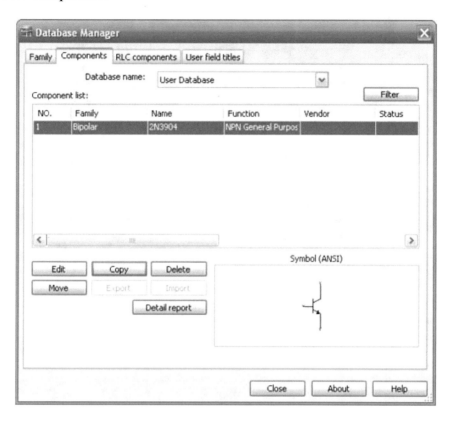

Figure 1.12: Elements in the User Database. Only the transistor 2N3904 is displayed.

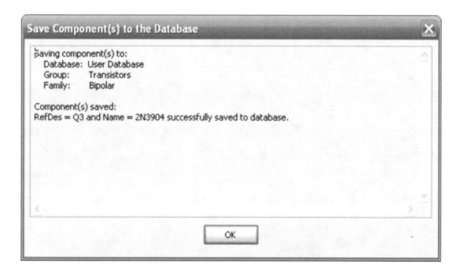

Figure 1.13: Message indicating the successful copy to the User Database.

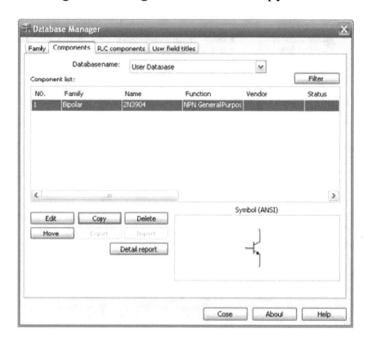

Figure 1.14: Database Manager with the Components tab for the User Database.

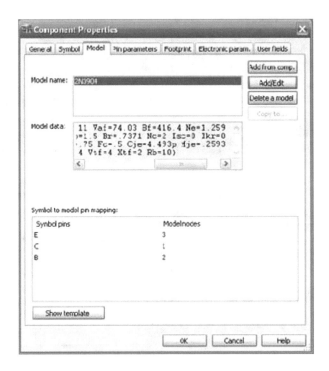

Figure 1.15: Component Properties for the selected component.

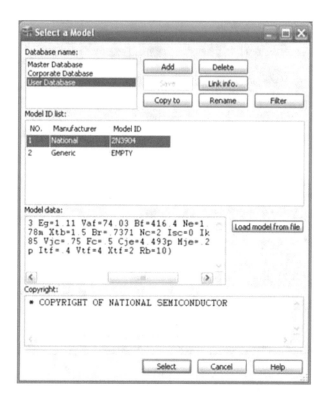

Figure 1.16: Dialog window to edit the model of the 2N3904.

We could also have renamed the component if after changing the desired parameters, in Figure 1.17 we press the Rename key and this takes us to the window shown in Figure 1.22 where we enter the new name.

Example 1.4 Common emitter amplifier with low β model.
We replace the BJT in the circuit of Example 1.1 with the transistor 2N3904_new (see Figure 1.23) and run the analysis to obtain the input and output waveforms of Figure 1.24. Here the smaller amplitude signal is the output. Measuring the gain we see that the new gain is now

$$Gain = -\frac{-4.262}{19.99} = -0.21$$

This is a considerable reduction in gain. The considerable loss in gain is due mainly to the absence of an emitter resistor which is used to reduce the effects of β variation in the transistor Q point [1].

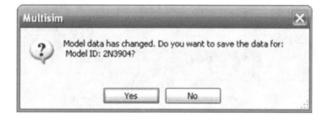

Select a Model

Database name:

Master Database
Corporate Database
User Database

Add   Delete
Save   Link info.
Copy to   Rename   Filter

Model ID list:

NO.	Manufacturer	Model ID
1	National	2N3904
2	Generic	EMPTY

Model data:

```
=1.11 Vaf=74.03 Bf=85.0 Ne=1.259
Xtb=1.5 Br=.7371 Nc=2 Isc=0 Ikr=0
jc=.75 Fc=.5 Cje=4.493p Mje=.2593
f=.4 Vtf=4 Xtf=2 Rb=10)
```

Load model from file

Copyright:

Select   Cancel   Help

Figure 1.17: Dialog window to edit the model of the 2N3904.

Multisim

? Model data has changed. Do you want to save the data for:
Model ID: 2N3904?

Yes   No

Figure 1.18: Dialog window to save the changes to the model.

Figure 1.19: Dialog window to overwrite the model after the changes are made.

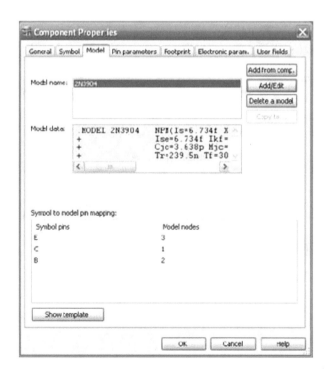

Figure 1.20: The parameter BF has been changed to 85 for the 2N3904.

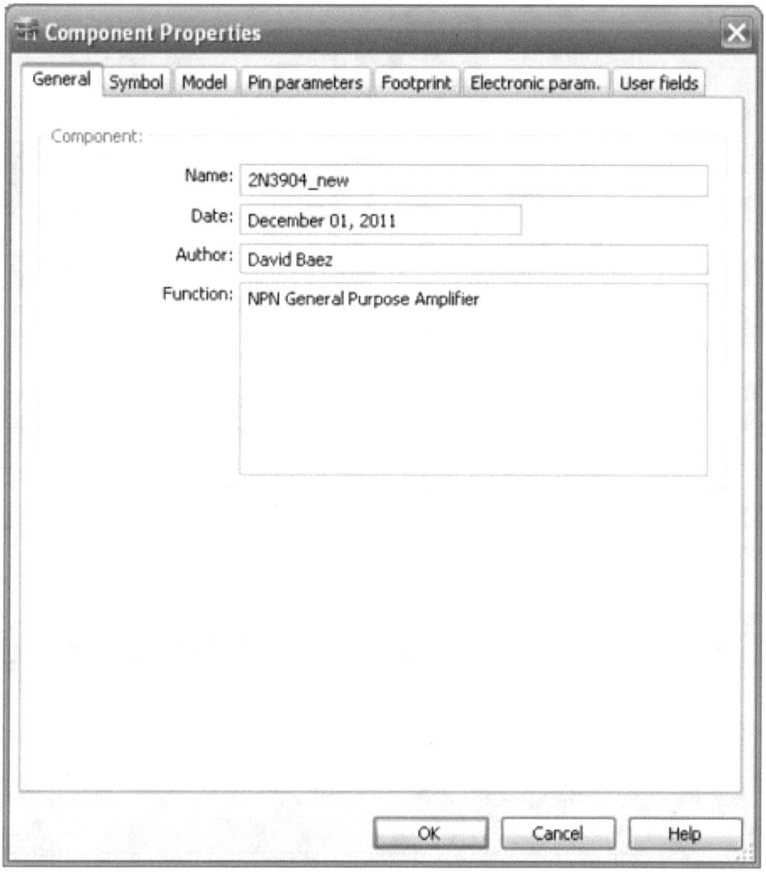

Figure 1.21: Component's model name has been changed to 2N3904_new.

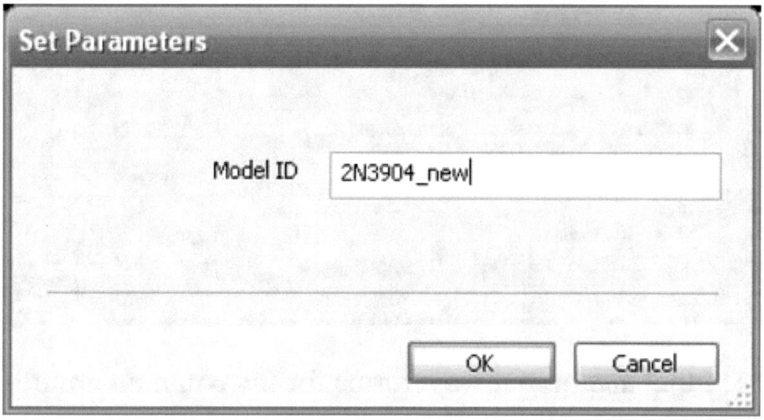

Figure 1.22: Component's model name has been changed to 2N3904_new.

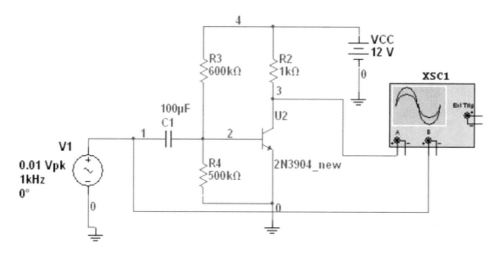

Figure 1.23: Common emitter amplifier with new transistor from the User Database 2N3904_new which has Bf = 54.

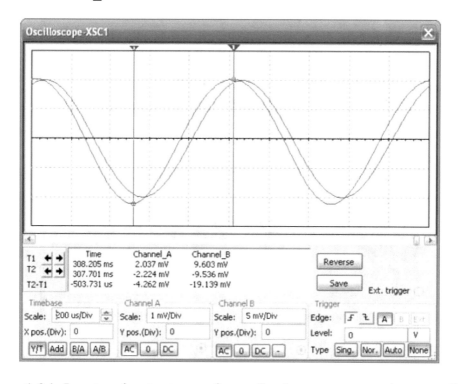

Figure 1.24: Input and output waveforms for the common emitter amplifier. The output voltage is much less than in the Example 1.1.

At this point, it is worth mentioning that we can completely create a new model according to requirements. We might use a model from a manufacturer. Manufacturer's models are usually available from manufacturer's web pages. If we need a new model, we just paste it in the Model Data space and continue the procedure.

1.2 CREATING SUBCIRCUITS IN MULTISIM

The use of subcircuits enhances the capabilities of circuit simulation. By using subcircuits we can create new components not provided in the databases. Subcircuits are useful when there is a circuit portion which repeats several times in a larger circuit. Thus, a larger circuit can be divided into smaller modules facilitating its diagram description. Multisim users can create subcircuits by using the Component Wizard provided in the main menu. There are two methods to create subcircuits. The first method creates a subcircuit to be used only in the circuit where it was created. In this case, a set of parts can be grouped together for reuse within the same circuit. The second method creates a subcircuit that can be kept in the Database for use in any other circuit simulation. In this case, we can create a part for a subcircuit provided by a vendor. We show all of these procedures with examples.

Example 1.5 Subcircuit for use within a circuit.
Let us consider the circuit of Figure 1.25. This is an RC-active KHN filter using three operational amplifiers [2]. The first operational amplifier is a summing amplifier while the last two are Miller integrators. The first integrator is formed by C2, R4, and U2. The second integrator is composed by C1, R6, and U3. If the output is taken from the first integrator's output, then the frequency response is that shown in Figure 1.26. It shows that the circuit realizes a band pass filter. We can group each of the integrators within a subcircuit. We do this by selecting the portion of the circuit as shown in Figure 1.27.

Now, from the main menu select Place→Replace by Subcircuit (Figure 1.28). This opens the dialog window shown in Figure 1.29 requesting a name for the subcircuit. After clicking on the OK button, Multisim asks if the merging of the nodes in the subcircuit and in the circuit is correct. We click on the OK button as many times as it is requested. When we are finished, the window shown in Figure 1.31 opens. Here we edit the subcircuit. We click on the Edit HB/SC button. The circuit inside the subcircuit block is open, as shown in Figure 1.32. We edit the circuit as needed, save it, and close the editor page. The edited circuit appears in Figure 1.33.

The final circuit using subcircuit blocks for both integrators in the filter is now ready for simulation. We perform an AC analysis to obtain the frequency response shown in Figure 1.35 which agrees with the one obtained in Figure 1.26

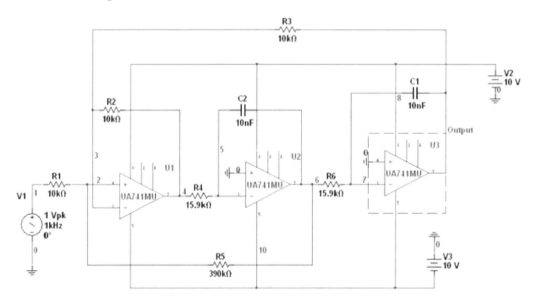

Figure 1.25: Active-RC KHN filter.

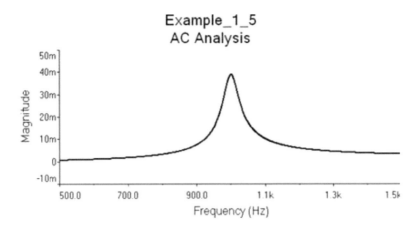

Figure 1.26: Frequency response showing the bandpass response.

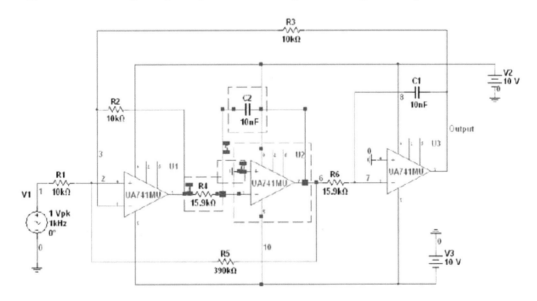

Figure 1.27: Portion of the filter selected for the subcircuit.

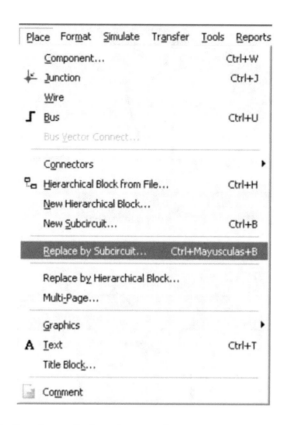

Figure 1.28: Path for the subcircuit creation.

Figure 1.29: We here give the name for the subcircuit.

Figure 1.30: Choosing the names for the subcircuit pins.

1.2.1 SUBCIRCUIT TO BE LOCATED IN THE DATABASE

In this section we present with an example the procedure to create a subcircuit from a node description of the circuit. We use the Component Wizard to create the subcircuit.

Example 1.6 Subcircuit for an operational amplifier macromodel available in the database. In this example we create a subcircuit block for the operational amplifier macromodel of Figure 1.36.

The circuit description in terms of the connections at nodes 1, 2, 3, Vin+, Vin-, and Vout is:

Rin VIN+ VIN- 2MEG
Epole 1 0 VIN+ VIN- 200000
Rpole 1 2 1k

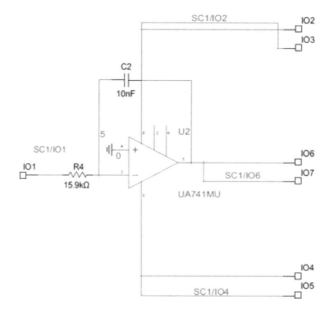

Figure 1.31: Dialog window to edit the subcircuit.

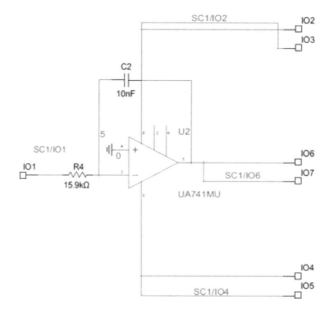

Figure 1.32: Schematic diagram for the subcircuit before deleting unneeded pins.

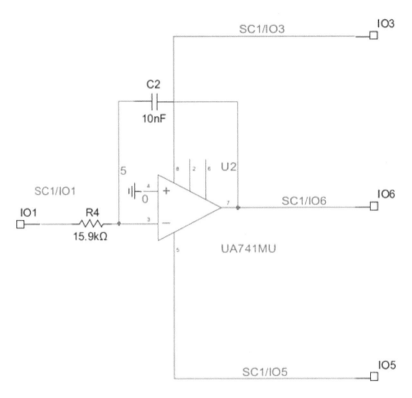

Figure 1.33: Subcircuit final schematic diagram.

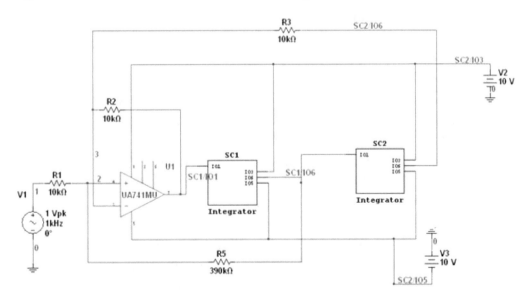

Figure 1.34: KHN filter with integrator subcircuits.

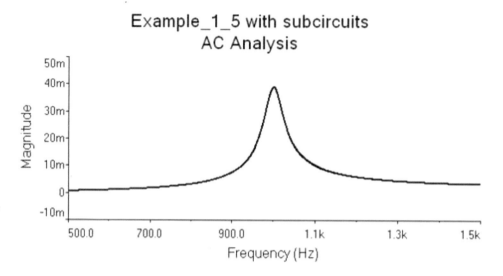

Figure 1.35: Frequency response for the filter using the subcircuit.

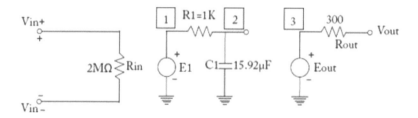

Figure 1.36: Macromodel for an operational amplifier.

Cpole 2 0 15.92u
Eout 3 0 2 0 1
Rout 3 OUT 300

where the component name is given first, the nodes where the component is connected follow, and finally the value. For the VCVS Epole, which describes the dominant pole, and Eout, which gives the low output impedance, we need, after the first two node names, the nodes where the VCVS takes is value, and finally the gain.

The procedure starts by selecting from the main menu Tools →Component Wizard which opens the dialog window of Figure 1.37. In this window we provide subcircuit information such as the component name, author's name, component type from among Analog, Digital, Verilog_HDL, or VHDL. Finally, we briefly describe the function realized by the subcircuit. We can see that this is the first window out of either 8, 7 or 6. The number of steps required in the subcircuit creation process depends upon the option we select in the radius buttons in this dialog window. For our example we select Simulation only (model) which needs 7 steps.

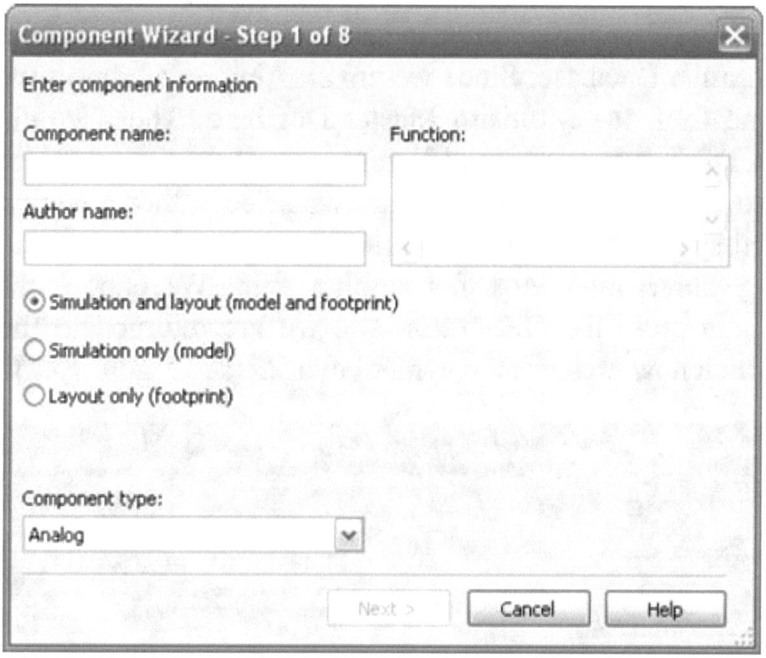

Figure 1.37: Dialog window for the Component Wizard.

We provide then the following information:

Component type name:	UDLAP_741
Author Name:	Baez_Guerrero_Cervantes
Component Type:	Analog
Function:	Op-amp single pole macromodel.

We select the radius button for: Simulation only (model). The window with the data is shown in Figure 1.38. We then press Next and obtain the second window for the Component Wizard, shown in Figure 1.39.

In this second window we select the radius button for: Single Section Component. This indicates that the part only has a macromodel. If the part has two or more macromodels, we select the button: Multi-Section Component. This is the case of TTL digital circuits where there are usually more than one circuit in the package. We also select 3 as the number of pins since we have two inputs and one output. Then, we click on the Next button. The window of Figure 1.40 is displayed now.

In this step we edit the subcircuit symbol. Although we can edit the symbol with the Symbol editor, it is better if we start from a known circuit symbol from the Database. Since we are creating an op amp subcircuit, we can use one from the Multisim Master Database. Thus, we click on the Copy from DB button to arrive at Figure 1.41. Here we can select a symbol by going to the appropriate library and we select either a component with the required symbol or a symbol looking closer to what we need. We select the Analog library and look for any op amp. We choose the op amp LM308M and press the OK button and we are returned to the previous window which now presents the symbol chosen (see Figure 1.42).

Figure 1.38: First dialog window for Component Wizard with data for the macromodel.

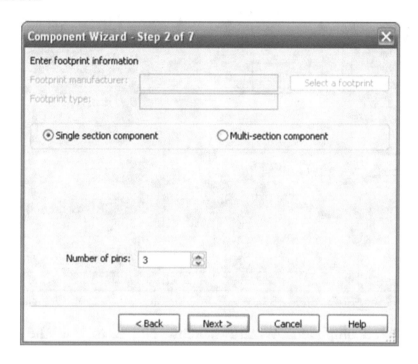

Figure 1.39: Second dialog window for Component Wizard.

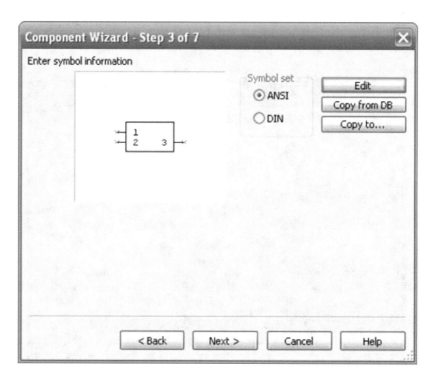

Figure 1.40: Component Wizard window No. 3 to edit the subcircuit symbol.

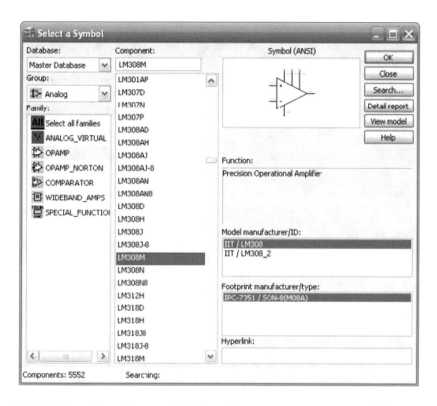

Figure 1.41: Choosing the symbol for the op amp macromodel.

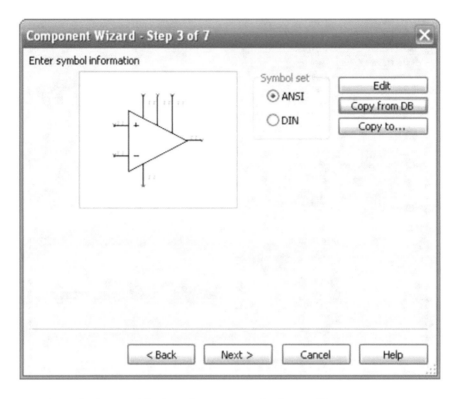

Figure 1.42: Window with the chosen symbol to edit.

We are now ready to edit the symbol to fit our requirements. We now press the Edit button. This takes us to the Symbol Editor shown in Figure 1.43.

The Symbol editor has toolbars which allows adding or deleting components in the circuit symbol. We can also delete rows from the spreadsheet at the bottom of the Symbol editor. After deleting the lines and pins that we do not need we get to the circuit symbol of Figure 1.44. We save the changes and close the Symbol Editor window to get to the window of Figure 1.45. If we are satisfied with the symbol, we press the Next button to arrive at Figure 1.46a where we select the pin function from the several options available for each pin, namely, bidirectional, unidirectional, input, output, and power. For the input pins we select INPUT and for the output pin we select OUTPUT as shown in Figure 1.46b.

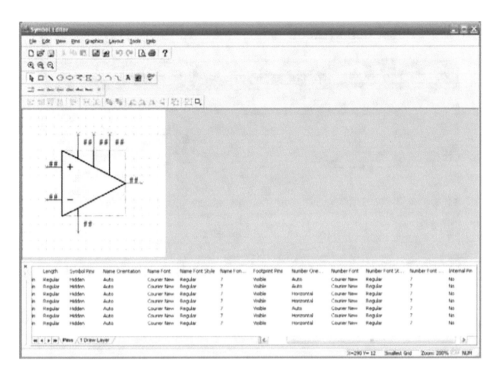

Figure 1.43: Symbol editor window.

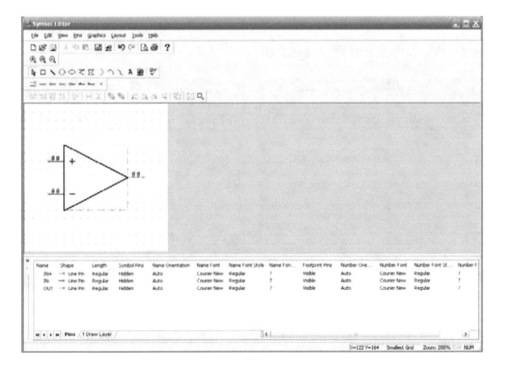

Figure 1.44: Subcircuit symbol.

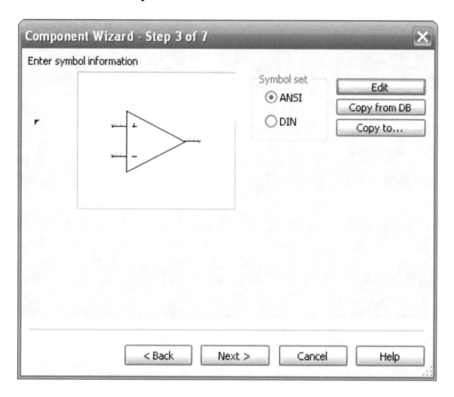

Figure 1.45: Subcircuit final symbol.

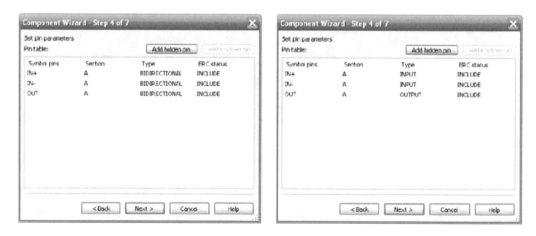

Figure 1.46: Selection of pin description.

Figure 1.47: Window to write the subcircuit description.

Pressing the Next button we get to step 5 of the Component Wizard shown in Figure 1.47. In this window we write the model name and the circuit description, as shown in Figure 1.48. We use the netlist that describes the macromodel and we have added the row:

.SUBCKT UDLAP_741 vin+ vin- out

This row indicates that the circuit description is a subcircuit with the name UDLAP_741. The three variables vin+, vin-, and out are the three pins to the subcircuit and they must always be given in this order. There are three rows that start with an asterisk. These are comment rows and do not affect the macromodel. There is also a row with .ENDS which indicates the end of the subcircuit description. When we are finished, we press the Next button to get to Figure 1.49 that describes the mapping information between the symbol pins and the model nodes. If we agree with that description, we click on the Next button.

Finally, we get to the Component Wizard window shown in Figure 1.50. In this window we select the Database; in our example we select the User Database. We also select the family; in this example we choose the

Analog group and add the family AO. This part of the process is similar to the one shown in Example 1.5. We click on the finish button and this ends the subcircuit creation procedure.

We can check the existence of the new part if we select the components in the User Database in the Analog group and the family AO, as shown in Figure 1.51.

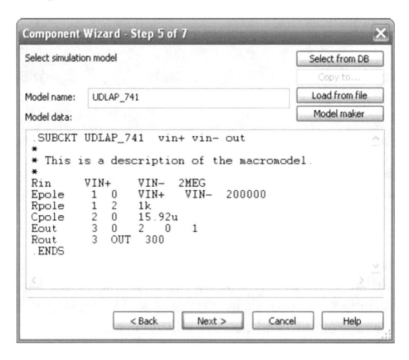

Figure 1.48: Subcircuit description for the Op amp UDLAP_741.

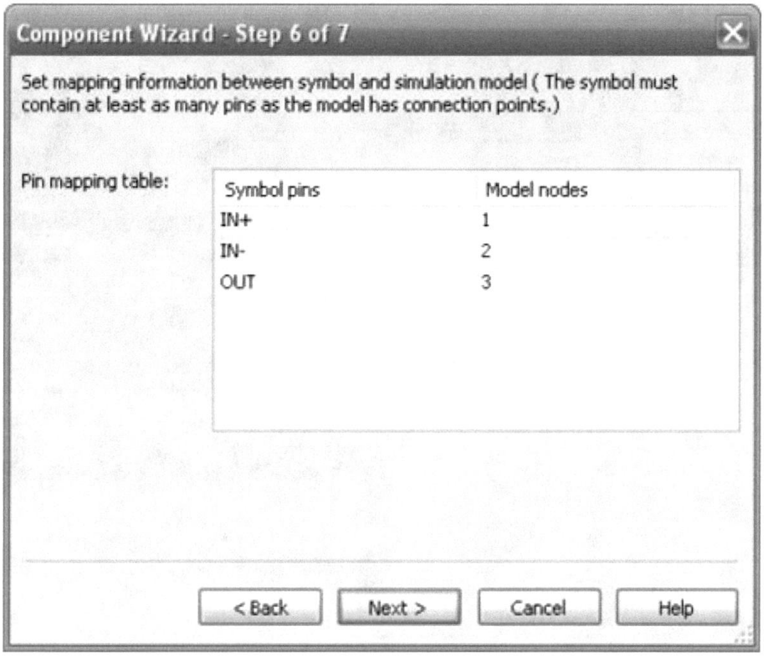

Figure 1.49: Pin assignment.

Figure 1.50: Subcircuit description.

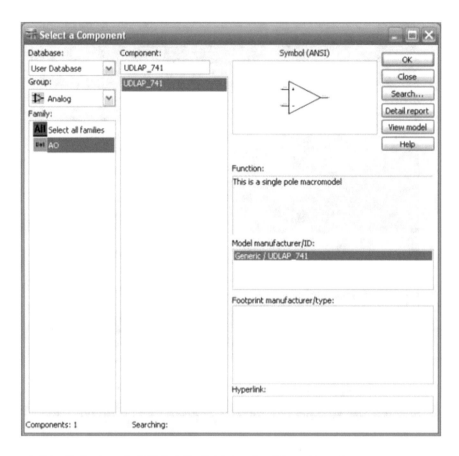

Figure 1.51: Subcircuit UDLAP_741 in the User Database.

Example 1.7 Inverting amplifier using the 741_UDLA op amp.

To test the macromodel, we wire up an inverting amplifier with a gain of -10 as shown in Figure 1.52. We perform a transient analysis from 0 to 2 msec. The variables we wish to plot are the input voltage V1 and the output voltage V3, as shown in Figure 1.53. After the transient analysis is done we obtain the plots of Figure 1.54 where the gain is -10, as expected.

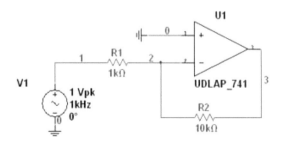

Figure 1.52: Inverting amplifier.

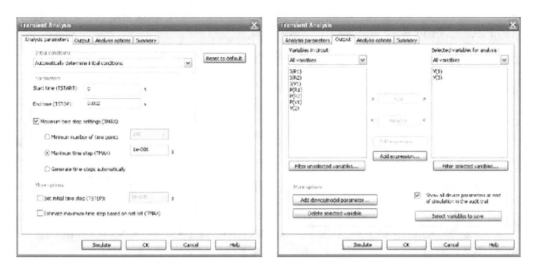

Figure 1.53: a) Transient analysis specifications and b) variables to plot.

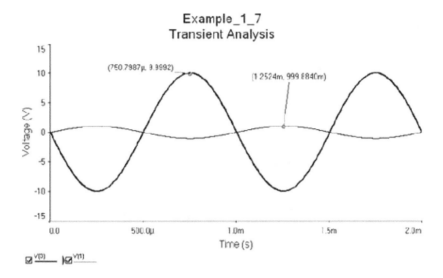

Figure 1.54: Inverting amplifier transient response.

1.3 CONCLUSIONS

An important part of the simulation is the modelling of each of the circuit components. In this chapter we have learned how to modify existing models, so they are suitable with a designer particular needs. We have also covered the way a subcircuit is created. Subcircuits are useful whenever either a circuit is repeated or it is going to be used as a module. We have also covered the way schematic symbols.

1.4 PROBLEMS

1.1. The inverter circuit in Figure 1.55 uses NMOS transistors. Use the model shown below for them. Test the circuit with a square signal. (The + sign at the beginning of a row a indicates continuation row.)

.MODEL CMOSN NMOS LEVEL = 2 PHI = 0.600000 TOX = 4.3500E-08

+ XJ = 0.2U TPG = 1 VTO = 0.8756 DELTA = 8.5650E+00 LD = 2.395E-07

+ KP = 4.5494E-05 UO = 573.1 UEXP = 1.5920E-01 UCRIT = 5.9160E+04

+ RSH = 1.0310E+01 GAMMA = 0.4179 NSUB = 3.3160E+15 NFS = 8.1800E+12

+ VMAX = 6.0280E+04 LAMBDA = 2.9330E-02 CGDO − 2.8518E-10

+ CGSO = 2.8518E-10 CGBO = 4.0921E-10 CJ = 1.0375E-04 MJ = 0.6604

+ CJSW = 2.1694E-10 MJSW = 0.178543 PB = 0.800000

* Weff = Wdrawn - Delta_W

* The suggested Delta_W is -4.0460E-07

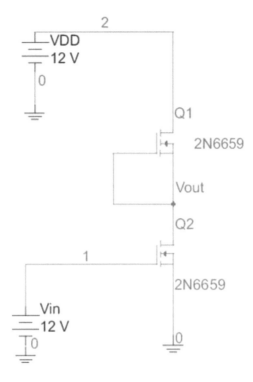

Figure 1.55: NMOS inverter circuit.

1.2. MOSIS provides models for diodes and transistors in different processes. The page is www.mosis.com. Test the inverter circuit in Example 1.1 with transistors with 0.5 microns process.

1.3. Test an inverter 7404 TTL by changing the rise and fall time to 10 msec.

1.4. The netlist below belongs to the TLC2201 op amp macromodel. Create a part for this circuit and store in a family group in the MISC group. Build a non-inverting amplifier with a gain of +10. Use 9KΩ and 1KΩ resistors.

* TLC2201 OPERATIONAL AMPLIFIER "MACROMODEL" SUBCIRCUIT

* CREATED USING PARTS RELEASE 4.03 ON 08/06/90 AT 15:18

* REV (N/A) SUPPLY VOLTAGE: 5V

* CONNECTIONS: NON-INVERTING INPUT

* | INVERTING INPUT

```
* | | POSITIVE POWER
* | | | NEGATIVE POWER SUPPLY
* | | | | OUTPUT
* | | | | |
.SUBCKT TLC2201 1 2 3 4 5
*
C1 11 12 11.00E-12
C2 6 7 50.00E-12
DC 5 53 DX
DE 54 5 DX
DLP 90 91 DX
DLN 92 90 DX
DP 4 3 DX
EGND 99 0 POLY(2) (3,0) (4,0) 0 .5 .5
FB 7 99 POLY(5) VB VC VE VLP VLN 0 537.9E3 -50E3 50E3 50E3
-50E3
GA 6 0 11 12 282.7E-6
GCM 0 6 10 99 2.303E-9
HLIM 90 0 VLIM 1K
ISS 3 10 DC 125.0E-6
J1 112 10 JX
J2 12 1 10 JX
R2 6 9 100.0E3
RD1 60 11 3.537E3
RD2 60 12 3.537E3
RO1 8 5 188
RO2 7 99 187
RP 3 4 5.71E3
RSS 10 99 1.600E6
VAD 60 4-.5
```

VB 9 0 DC 0
VC 3 53 DC .928
VE 54 4 DC .728
VLIM 7 8 DC 0
VLP 91 0 DC 2.800
VLN 0 92 DC 2.800
.MODEL DX D(IS=800.0E-18)
.MODEL JX PJF(IS=500.0E-15 BETA=1.279E-3 VTO=-.177)
.ENDS

REFERENCES

[1] A.S. Sedra and K.C. Smith, Microelectronic Circuits, Holt, Rinehart, and Winston, New York, 2005. Cited on page(s) 12

[2] L.P. Huelsman, Introduction to the Theory and Design of Active and Passive Filters, McGraw-Hill Co., New York, 1991. Cited on page(s) 17

CHAPTER 2

Transmission Lines

2.1 INTRODUCTION

Transmission lines are parts of a circuit that interconnect the different components of a circuit. They are also used to model transmission lines in communication systems. In many integrated circuits there are losses due to the interconnections between the different components. These losses are due to limitations such as the bandwidth, attenuation, and distortion, among many other factors, which in many cases are negligible because the circuits operate at low frequencies. An interconnection consisting of two conductors and a dielectric can be considered as a transmission line. Transmission lines have distributed parameters and have a characteristic impedance which is a dynamic parameter defined by the ratio of voltage and current in any given point along the transmission line. It is denoted by Z0 and can be expressed in terms of the distributed inductance and the capacitance of the transmission line. The propagation delay is also a characteristic of the transmission line and depends upon the same parameters. Transmission lines can be used to study impedance coupling effects. An interesting application of transmission lines occurs when the delay is used in the simulation of the delay of digital filters. An example shows this application of transmission lines. Multisim has available lossless transmission lines as well as lossy ones.

2.2 PARAMETERS OF LOSSLESS TRANSMISSION LINES

Transmission lines are available from the Miscellaneous (MISC) set of components. There are two types of transmission lines. Type 1 transmission lines have as parameters the nominal impedance Z0 and the propagation delay TD. Type 2 transmission lines have as parameters the nominal impedance Z0, the frequency F, and the normalized electrical length NL. The schematic symbol for a transmission line is given in Figure 2.1 and its name begins with the letter W.

The corresponding dialog windows to specify the parameters of transmission lines are shown in Figure 2.2.

2.3 EXAMPLES

In this section we present four examples using lossless transmission lines in Multisim. The last example shows the use of transmission lines in the simulation of digital filters.

Example 2.1   Transmission line excited by a pulse.
As our first example in using transmission lines we have the circuit of Figure 2.3, where we analyze a circuit with a 50 ohms Type 1 transmission line with an input voltage V1 = 10 v, a period of 4 nsec, and a duty cycle of 25%; that is, the pulse width is 1 nsec. The parameters for W1 are given in Figure 2.4 which is opened by double clicking on the transmission line W1.

(a)   (b)

Figure 2.1: Symbol of a transmission line. a) Type 1, b) Type 2.

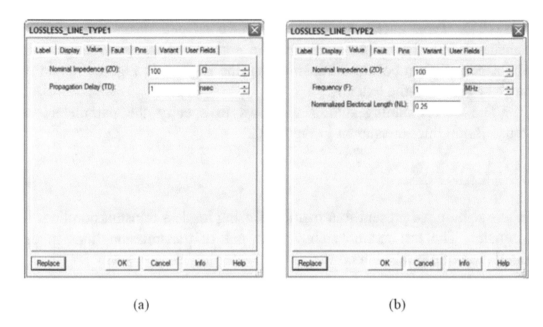

(a)   (b)

Figure 2.2: Dialog windows to specify the parameters of transmission lines. a) Type 1, b) Type 2.

In this example, the transmission line has a delay of 2 nsec, so we expect the signal to arrive at the output 2 nsec later of the input. We run a transient analysis with a TSTOP of 4 nsec. Plots of input and output signals are shown in Figure 2.5. We can see there that the output signal is delayed by 2 nsec when it traveled through the transmission line. Furthermore, the transmission line W1 is perfectly balanced because the terminating resistances have the same values as the impedance Z0 of the line. This makes that the amplitude of the output voltage be half of the amplitude of the voltage of the input signal.

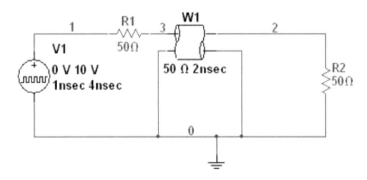

Figure 2.3: Circuit with a Type 1 transmission line.

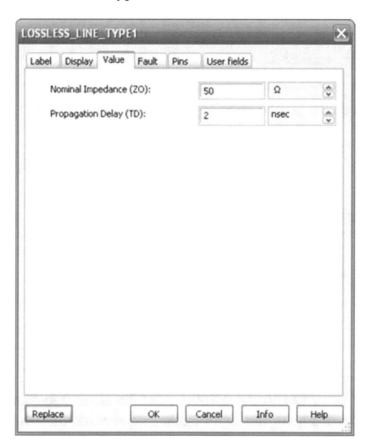

Figure 2.4: Parameters of the transmission line.

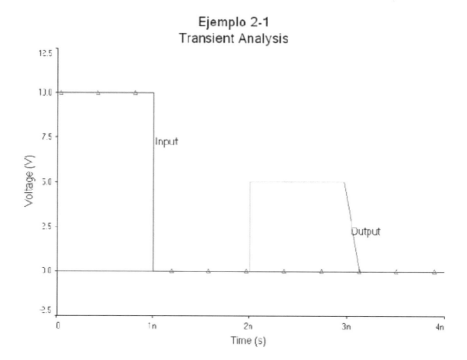

Figure 2.5: Input and output signals for R1 = R2 = Z0 =50 Ω.

Let us consider the case when resistors R1 and R2 are not equal valued. If R1 = 10 Ω and R2 = 100 Ω, for the same transient analysis we obtain he plots of Figure 2.6. In this case, we see that the output voltage is larger now than in the case of a balanced transmission line.

Example 2.2 Pulse with very small duty cycle.
Let us consider the case when the pulse has a very small duty cycle as compared to the delay of the transmission line. First we analyze the circuit when both terminating resistors are equal valued and their value is equal to Z0; that is, R1= R2= Z0= 50 Ω. The pulse width is 0.5 nsec with period 20 nsec. The value of TD is 2 nsec. The circuit is shown in Figure 2.7 and the result of the simulation is shown in Figure 2.8. The magnitude of the output voltage has the same value as in the previous example and it is a half of the input voltage.

Now we change the values of the terminating resistors with both values different from the impedance of W1; that is, they are not coupled.

Let R1 = 5 Ω and R2 = 30 Ω as shown in Figure 2.9. Running a transient analysis we obtain the output plot of Figure 2.10.

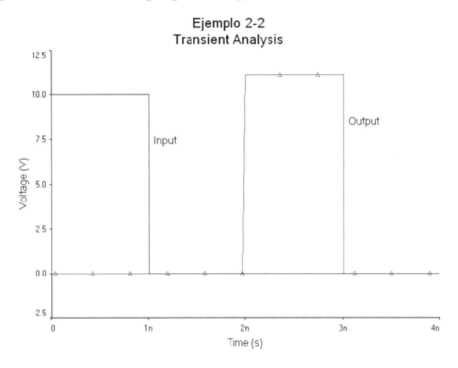

Figure 2.6: Input and output signals for R1 = 10 Ω, R2 = 100 Ω, Z0=50 Ω.

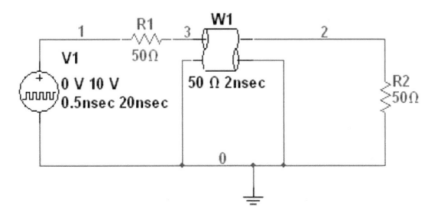

Figure 2.7: Pulse with small duty cycle shorter than the transmission line delay and R1 = R2 = Z0 = 50 Ω.

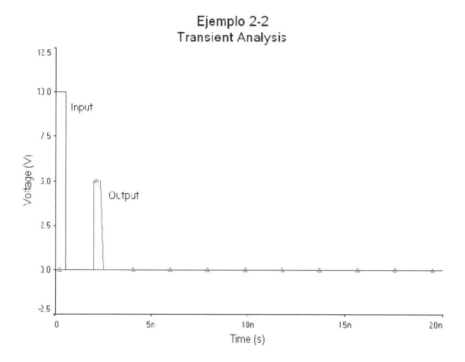

Figure 2.8: Input and output signals when the duty cycle is very small.

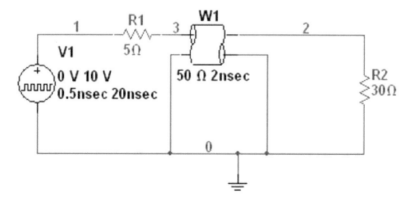

Figure 2.9: Circuit with impedances decoupled and very small duty cycle.

Figure 2.10 shows a series of reflections that occur inside the transmission line and they are due to the mismatch that exists among the impedance of the transmission line W1 and the terminating resistors R1 and R2. The first output pulse occurs 2 nsec after the input is applied to the line whereas the second output pulse occurs 4 nsec after the first output pulse because part of the output signal is reflected back to the input where it is again reflected to the output. Part of this pulse appears at the output and part is reflected back to the input where the process is repeated again, each time with less amplitude as we can see in Figure 2.10. There we can still appreciate a very small pulse at 14 nsec.

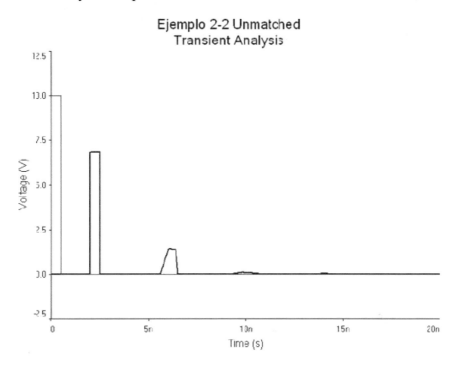

Figure 2.10: Input and output signals when the impedances are not coupled. Note the reflections in the output signal due to the decoupling and the very small duty cycle.

Example 2.3 Transmission line coupling.
In this example we make an analysis of a circuit to couple a load using transmission lines [3]. Figure 2.11 shows a circuit where transmission lines W1, W2, and W3 transform the 50 ohms impedance of R1 to 5 ohms, which is a very low load resistance. The parameters for the transmission lines are shown Figure 2.12. An AC analysis form 0 to 100 MHz produces the frequency response of Figure 2.13.

Example 2.4 Passband digital filter.

As mentioned above, an important characteristic of transmission lines is the propagation delay TD. This characteristic allows us to use transmission lines to simulate digital filters [1] since delays are a basic component of digital filters. A biquadratic digital filter topology is shown in Figure 2.14a. The basic block is shown in Figure 2.14b. The variable z^{-1} represents a delay of the signal by time T, known as sampling period. This delay can be simulated by the transmission line delay by making T = TD.

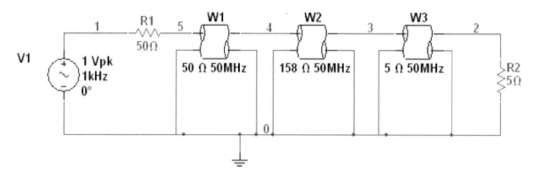

Figure 2.11: Load coupling circuit.

Figure 2.12: Parameters for the AC analysis.

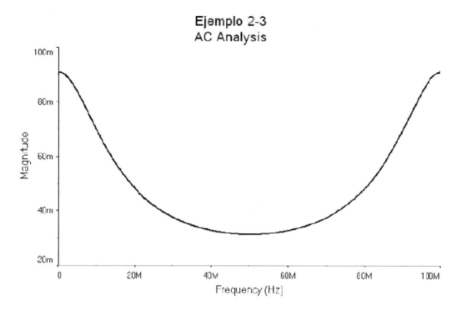

Figure 2.13: Frequency response of the impedance coupling circuit.

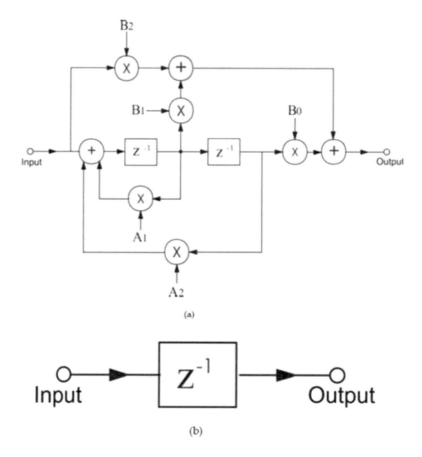

(a)

(b)

Figure 2.14: a) Biquadratic digital filter, b) Unit delay.

The transfer function of a biquadratic second order digital filter can be written as

$$H(z) = \frac{B_0 z^2 + B_1 z + B_2}{z^2 + A_1 z + A_2}$$

The digital filter can be realized in Multisim with the circuit of Figure 2.15. In this circuit W1 and W2 emulate the delays; VCVS V2 and V3 simulate ideal operational amplifiers to realize the adders and the multipliers. V2 together with resistors R1, R6, R4, and R2a form a summing amplifier. V3 together with R10, R11a, R6, and R8 form a second inverting summing amplifier. VCVS V1 and V4 are unity gain buffers to feed the signal to the other stages. VCVS V7 together with R5 realize the negative coefficient B1, while V6 together with R8 realize the B0 coefficient. Coefficients B_0, B_1, B_2, A_1 and A_2, determine what type of filter is realized. These coefficients can be obtained using software for digital filter design such as Winfilters [2]. The circuit we wish to realize is a second order passband Butterworth filter with a 3 dB passband from 900 Hz to 1100 Hz. The sampling frequency is 6 KHz which is equivalent to a sampling period of T = 1/6000 = 166.67 μ sec. Using Winfiltros we obtain the transfer function

$$H(z) = \frac{z^2 - 1}{z^2 - 0.9096707z + 0.809374}$$

Then the parameters of the transmission lines are

$Z0 = 1\ \Omega$ y $TD = 166.67\ \mu S$

Where TD is the sampling period of the digital filter. We do an AC analysis form 10 Hz to 2 KHz to obtain the frequency response shown in Figure 2.16. It is known that digital filters must obey Nyquist theorem; that is, they can only process frequencies up to half the sampling frequency. If we make an AC analysis beyond 3 KHz, which is half the sampling frequency we obtain the response of Figure 2.17.

Example 2.5 Lossy transmission lines.

The lossless transmission line model is adequate for short length transmission lines, but for long transmission lines the lossy model is more adequate. A lossy transmission line is modelled with the parameters shown in Table 2.1.

Some important parameters for transmission lines are the characteristic impedance Z_0 and the propagation constant γ. They are given by the following:

$$Z_0 = \sqrt{\frac{R + j\omega L}{G + j\omega C}}$$

Example 2.6 Characteristic impedance of a lossy transmission line.

The characteristic impedance for the parameters for a lossy transmission line with a length of 1 km, $R = 15\ \Omega$, $C = 70\text{nF}$, $L = 175\ \mu\text{H}$, $G = 9\mu\text{S}$, at a frequency $f = 2$ MHz, we have $\omega = 2\pi f = 2\times\pi\times2\times10^6 = 12.6$ Mrad/sec.

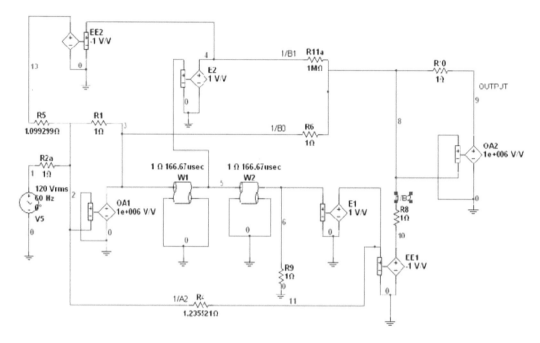

Figure 2.15: Digital filter using transmission lines.

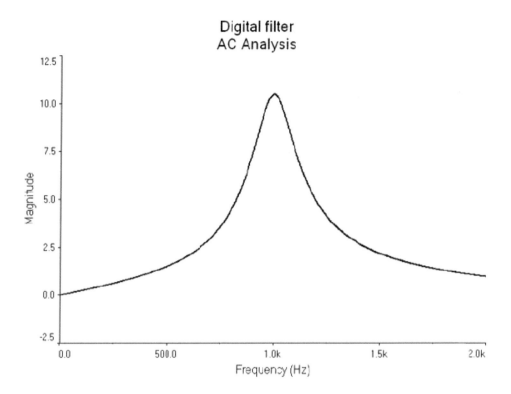

Figure 2.16: Frequency response of the digital filter.

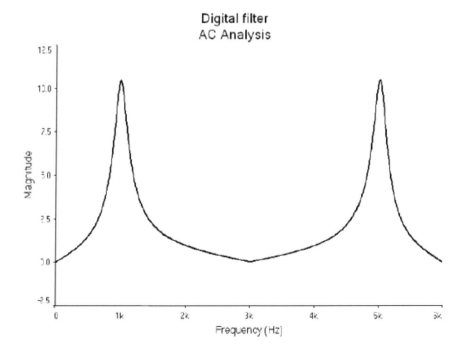

Figure 2.17: Frequency response of the digital filter after an AC analysis up to 6 KHz.

Table 2.1: Parameters of a lossy transmission line.

Parameter	Physical characteristic	Unit	Example
LEN	Electrical length	Meter	1 km
R	Resistance/electrical length	Ω/LEN	10Ω/km
L	Inductance/electrical length	Henry/LEN	10mH/km
G	Conductance/electrical length	Siemen/LEN	10 mS/km
C	Capacitance/electrical length	Farad/LEN	10 μF/m

$$Z_0 = \sqrt{\frac{R + j\omega L}{G + j\omega C}} = \sqrt{\frac{15 + j(12.6 \times 10^6)(175 \times 10^{-6})}{9 \times 10^{-6} + j(12.6 \times 10^6)(70 \times 10^{-9})}} = 35.4756 + 35.2351i$$

Example 2.7   Lossy transmission line.

Let us consider the circuit in Figure 2.18. It has the parameters of the previous example. These parameters are shown in Figure 2.19. The length is normalized to unity. We note that from the previous example there is an impedance mismatch between the real part of the characteristic impedance and resistor R1. We perform an AC analysis and measure the input impedance given by the voltage at node 3 divided by the current passing through resistance R1, that is, we wish to plot V(3)/I/R1). The analysis setup is shown in Figures 2.20 a and b. The resulting plot is shown in Figure 2.21. With the cursor we see that at almost 2 MHz the input impedance is 50.1 Ω.

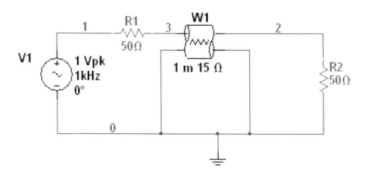

Figure 2.18: Circuit with a lossy transmission line.

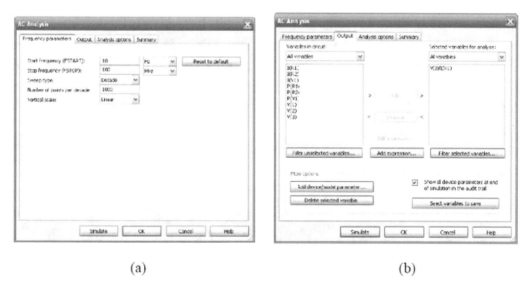

Figure 2.19: Parameters of the lossy transmission line.

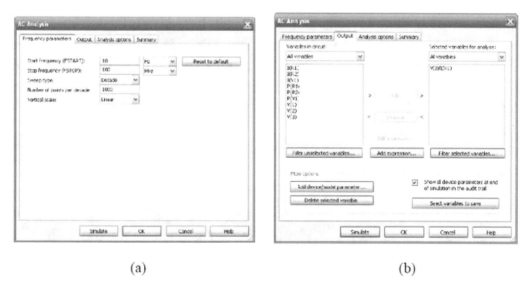

(a) (b)

Figure 2.20: Analysis setup. a) AC analysis parameters, b) function to plot.

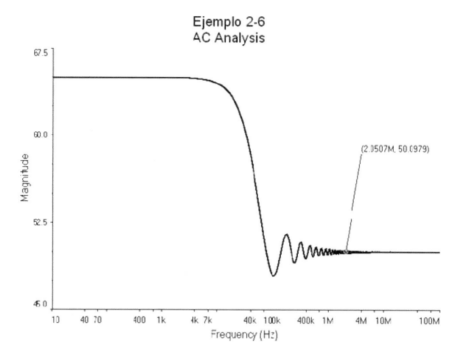

Figure 2.21: Measurement of the lossy transmission line.

2.4 CONCLUSIONS

In this chapter we have introduced transmission lines. Transmission lines appear in integrated circuit, printed circuit boards, and communication systems. The examples presented here show the use of transmission lines. Multisim has two types of transmission lines denoted by Types 1 and 2. They are equivalent and the only difference is the set of parameters that define them. The use of the transmission delay is used to simulate digital filters.

2.5 PROBLEMS

2.1. In the circuit of Figure 2.3 a lossless transmission line with TD = 2 μs, and Z0 = 50 Ω, apply a signal Piecewise linear with the following parameters:

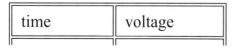

time	voltage

0	0
1n	100
6u	100
6.001u	0
50u	0

Use R1 = 150 and R2 = 1u. Plot the voltage measured at the input of the transmission line.

2.2. Repeat Exercise 2.1 but with R1 = $1\mu\Omega$ and R2 = 150 Ω. Repeat with an input signal defined by

time	voltage
0	0
0.001u	30

and plot the voltage at the output of the transmission line.

2.3. Obtain the characteristic impedance for a lossy transmission line with R = 2.25 Ω/m, L = 1μH/m, C = 100 pF/m, and G = 0 at 500 MHz. Using Multisim plot the input impedance if the load is R = infinite and if R = 10.

2.4. In the circuit of Figure 2.18 use LEN = 20, R = 10, L = 100u, G = 10u, and C = 70n. Apply a square signal with a period of 0.8ms and a 50% duty-cycle with voltage levels 0 and 1 V. Observe and explain the low-pass behavior.

2.5. In the circuit in Example 2.1 change R2 to a) Infinity, b) 0, c) 200, and d) 25.

2.6. Repeat Exercise 2.5 for Example 2.6.

REFERENCES

[1] H. Nielinger, Digital (IIR) Filter Biquad Section Simulated with PSpice, IEEE Trans. on Education, Nov. 1993, Vol. 36, No. 4, pp 383-385. Cited on page(s) 45, 69

[2] D. Báez-López et al, Multimedia Based Analog and Digital Filter Design, Computer Applications in Engineering Education, pp.1-8, No.1, vol. 6, 1998. Cited on page(s) 48

[3] P. R. Clayton, Analysis of Multiconductor Transmission Lines, 2nd Edition, J. Wiley, NY, 2007. Cited on page(s) 45

CHAPTER 3

Other Types of Analyses

INTRODUCTION

This chapter will cover some types of analyses that will complement and extend the basic analysis DC, AC, and transient analyses. First, it describes sensitivity analysis. It then treats noise analysis which can be done only if an AC analysis is also performed. Then, it describes how to perform Monte Carlo and Worst-case analyses. Finally, it describes how to perform parametric and temperature analyses.

3.1 SENSITIVITY ANALYSIS

An analysis which is of considerable importance in the design of circuits is sensitivity analysis. There are several definitions for sensitivity. Multisim evaluates the sensitivity of a node voltage or a branch current F(x) with respect to a circuit element x as

$$S_x^{F(x)} = \frac{\partial F(x)}{\partial x}$$

and it has units of volts (amperes) per unit value. Sensitivities are only evaluated for DC and AC analyses. Sensitivity analysis is performed with respect to all elements of the circuit, whether active or passive. The steps carried out for a sensitivity analysis are as follows:

1. First either a DC or AC analysis is performed.

2. Then the circuit is linearized around the bias point and all the derivatives of the selected node voltages and branch currents are

calculated with respect to each and every one of the elements in the circuit.

The output includes the names of the elements, their values, the sensitivity in volts/unit or amps/unity. We show the procedure by means of an example.

Example 3.1 Voltage divider resistive circuit.
Let us consider the circuit of Figure 3.1. It is desired to perform a sensitivity analysis for this circuit. From the main menu we select Simulate →Analyses →Sensitivity to open the window in Figure 3.2 where we enter the value V(2) in the space for the voltage output node as shown and we also select the radius button for DC Sensitivity. Then we select the Output tab and we select all the variables as shown in Figure 3.3.

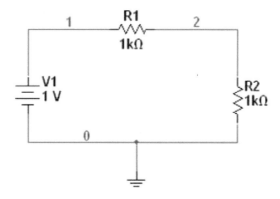

Figure 3.1: Resistive voltage divider circuit.

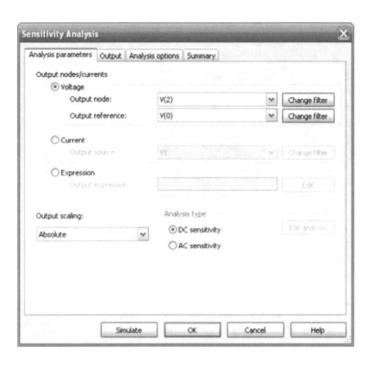

Figure 3.2: Dialog window to specify the output voltage variable for sensitivity analysis.

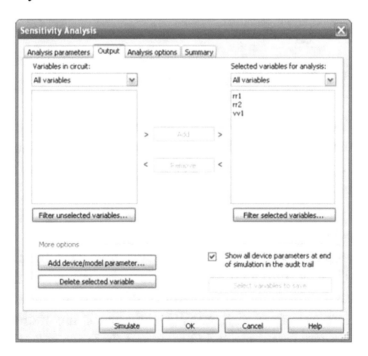

Figure 3.3: Selection of the variables for the sensitivity analysis.

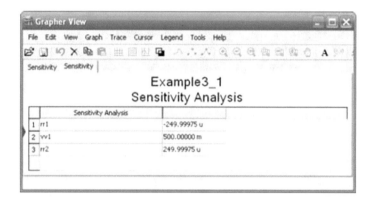

Figure 3.4: Results for the DC sensitivity analysis.

The results are shown in the Grapher window shown in Figure 3.4 and they indicate that the sensitivities calculated are:

$$S_{V1}^{V(2)} = \frac{\partial V(2)}{\partial V1} = -500 \times 10^{-3} \ volts/volt = -0.5 \ volts/volt$$

$$S_{R1}^{V(2)} = \frac{\partial V(2)}{\partial R1} = -250 \times 10^{-6} volts/ohm = -0.25 \ mV/ohm$$

$$S_{R2}^{V(2)} = \frac{\partial V(2)}{\partial R2} = +250 \times 10^{-6} volts/ohm = +0.25 \ mV/ohm$$

This information tells us that when R1 increases by 1 ohm, the voltage V(2) changes decreases 0.25 mV (it decreases because of the minus sign). Also that when R2 changes by 1 ohm, V(2) changes +0.25 mV, and that when V1 changes 1 volt, then V(2) changes -0.5 volts.

Example 3.2 Low pass filter.
The circuit shown in Figure 3.5 is a passive low pass filter. We wish to perform an AC sensitivity analysis for node V(2). We select from the main menu Simulate → Analyses → Sensitivity to obtain the dialog window of Figure 3.6. Here we enter voltage V(2) and mark the radius button AC Sensitivity. Then we select the tab Output and there we select the capacitor

C2 as shown in Figure 3.7. After clicking on the Simulate button we obtain the plot of Figure 3.8. We note that the sensitivity has a maximum around the 3 dB cut-off frequency which is 159 Hz.

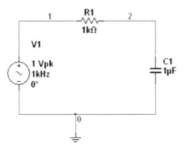

Figure 3.5: Low pass passive filter.

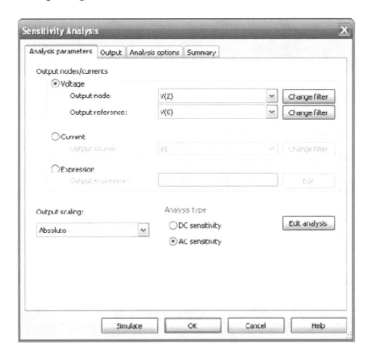

Figure 3.6: Dialog window to specify output node voltage for AC sensitivity analysis.

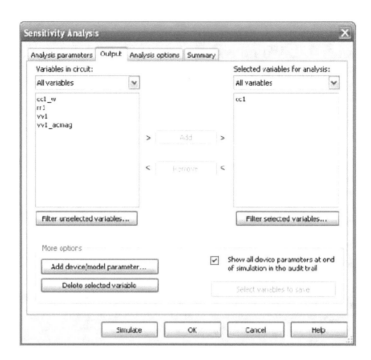

Figure 3.7: Selection of the element with respect to which sensitivity is taken.

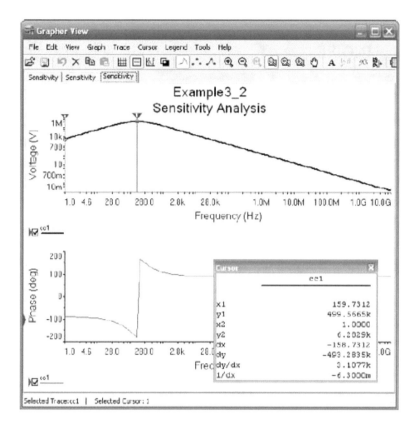

Figure 3.8: Plot of sensitivity of V(2) with respect to C1.

3.2 NOISE ANALYSIS

Noise analysis is performed in conjunction with an AC analysis. Noise analysis calculates the noise in the output voltage of a circuit due to the noise generated by devices such as resistors and semiconductor devices. To perform this analysis, Multisim generates a noise spectrum density for each device in a range of frequencies and perform an RMS sum in the output node. To make a noise analysis, from the main menu select Simulate $\rightarrow$ Analyses $\rightarrow$ Noise analysis and then enter the necessary output node, the frequency range, and the output noise. The noise densities obtained are the ONOISE and INOISE. They are the power spectral noise densities due to resistors and transistors in the circuit and the units are volts^2/Hz or amperes^2/Hz. ONOISE is the output referred noise and INOISE is the input referred noise. The following example shows the procedure to plot ONOISE and INOISE.

Example 3.3   Common emitter circuit.

To illustrate the noise analysis procedure, let us consider the common emitter amplifier of Figure 3.9. In this circuit, we wish to plot the input and output referred noise. For the analysis we choose node labeled as output as the output node. From the main menu we select Simulate → Analyses → Noise analysis to open the dialog window of Figure 3.10 where we enter the output node as V(Output). Then we select the tab for Frequency parameters and enter the frequency values as shown in Figure 3.11. Finally, we select in the Output tab the variables we wish to plot. In this example we choose the total INOISE and ONOISE as shown in Figure 3.12. We then click on the Simulate button to obtain the spectra noise plots shown in Figure 3.13.

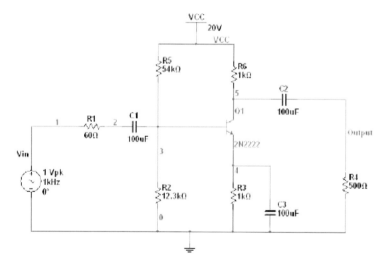

Figure 3.9: Common emitter amplifier.

Figure 3.10: Dialog window to select the output node.

In addition to the spectrum noise analysis, Multisim (only the Power Pro version) also calculates the Noise figure NF which is defined by

$$NF = 10 \log_{10} \left(\frac{SNR_{in}}{SNR_{out}} \right)$$

where SNR$_{in}$ and SNR$_{out}$ are the signal-to-noise ratios measured at the input and at the output, respectively. The Noise figure analysis can be done by selecting from the main menu Simulate $\rightarrow$ Analyses $\rightarrow$ Noise Figure Analysis. This opens the dialog window shown in Figure 3.14. There we select the input voltage source, which in this example is V3 and then we click on the Simulate button to obtain Figure 3.15 showing the value for the noise figure in dB with a value for the common emitter amplifier of NF = 34.5129 dB.

3.3 MONTE CARLO AND WORST CASE ANALYSES

A Monte Carlo analysis is a statistical analysis that allows us to observe the way component tolerances affect the circuit. There are two types of analyses that can be performed. Worst case analysis and Monte Carlo analysis.

Worst case analysis is used to calculate the worst value of a circuit parameter of interest for the given tolerances of the components. In this case the components of interest are varied to their maximum and minimum values according to the defined tolerances. On the other side, a Monte Carlo analysis calculates the response of the circuit when the values of the components vary randomly (only for the components for which wc specify a tolerance). In this case, several runs are performed using these tolerances. The difference between the two types of analyses is that while the worst case analysis shows that not all designs meet specifications, a Monte Carlo analysis shows which percentage of the circuits meet specifications.

3.3.1 MONTE CARLO ANALYSIS

Multisim begins conducting the analysis indicated with the values of all the elements and parameters in their nominal value. The results of this analysis are saved for later comparison with the analyses which are obtained by varying the values of the elements sclcctcd with a tolerance. To illustrate a Monte Carlo analysis let us consider an example.

Example 3.4   RC lowpass circuit.
Consider the RC passive low pass circuit excited by an AC voltage source, as shown in Figure 3.16. We have assigned a 10% tolerance to each passive component. From the main menu select Simulate →Analyses →Monte Carlo which opens the dialog window of Figure 3.17 where we assign tolerances to the components we are interested in having their values varied in the Monte Carlo analysis. We assign tolerance to each component by clicking on the Edit selected tolerance button which in turn opens the dialog window of Figure 3.18 where we assign tolerance to the component and we also choose which type of distribution it has from the two available choices either Gaussian or uniform distribution.

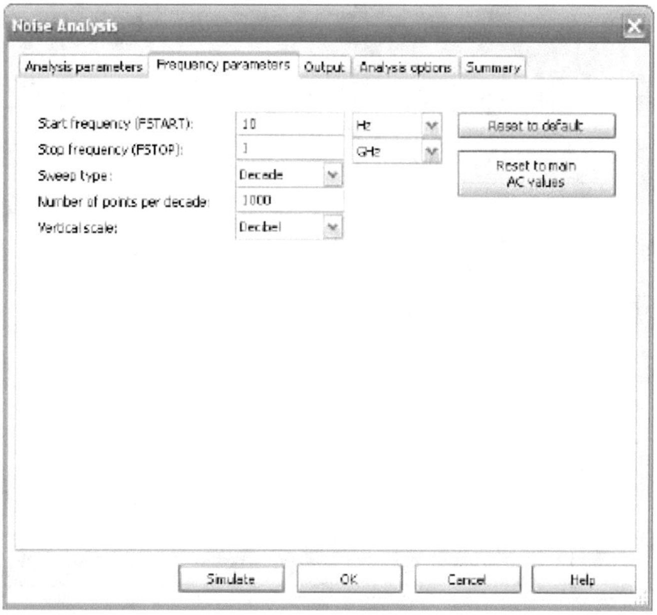

Figure 3.11: Dialog window to select the frequency range.

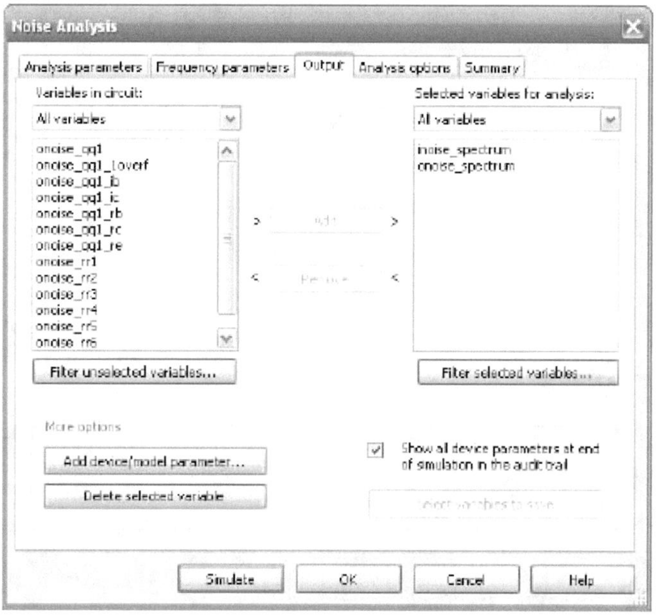

Figure 3.12: Selection of the noise densities.

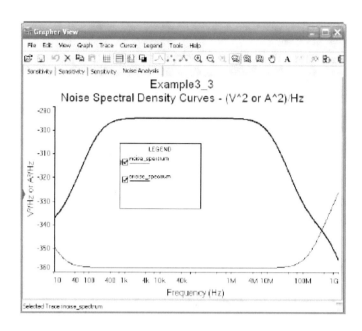

Figure 3.13: INOISE and ONOISE spectral densities.

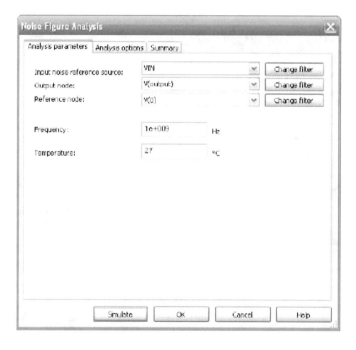

Figure 3.14: Dialog window to select the input voltage source for the Noise figure analysis.

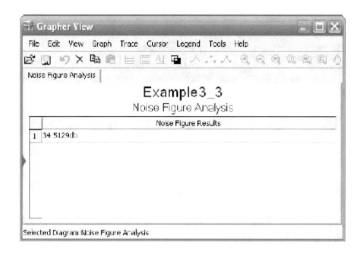

Figure 3.15: Noise figure for the common emitter amplifier.

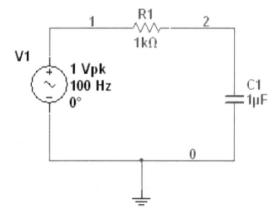

Figure 3.16: RC lowpass passive circuit.

Figure 3.17: Adding a tolerance to capacitor C1.

For the capacitor we assign 20% tolerance with a Gaussian distribution and then press Accept. We repeat the procedure for the resistor but apply a uniform distribution. After doing this we obtain the window of Figure 3.19 which shows the tolerance and distribution for each and every component with a tolerance assigned. To end the tolerance assignment we click on the OK button.

Now we select the Analysis Parameters tab as in Figure 3.20. Here we specify the analysis type from among DC Operating Point, AC Analysis, and Transient Analysis. For this example we choose a transient analysis. First we edit the transient analysis. With a click on the Edit Analysis button we open the window of Figure 3.21 where we give the data for the analysis. Since the voltage source has a 100 Hz frequency, we choose 0.012 sec as the stop time in the transient analysis and give a maximum step size of 1E-5 sec. After finishing entering the data for the analysis we press the OK button and return to the window of Figure 3.20. There we enter the data for the Monte Carlo analysis. We select the output variable which in this case is V(2). We also select the number of runs as 5 and the collating function. The more runs specified in the dialog window of a Monte Carlo analysis, the longer the time it takes to complete the simulation. We choose the collating function from among the ones available at Table 3.1 choosing MAX. We also select to plot all the responses in the same plot. Finally, we press the Simulate button and obtain the plots of Figure 3.22. There we see the nominal run and the other 5 runs with component values varied according to the tolerance and distribution chosen for the components. In Figure 3.23 we see a spreadsheet with the data of the different runs.

Figure 3.18: Dialog window to specify tolerance to capacitor C1.

Figure 3.19: Window showing component tolerances and distributions.

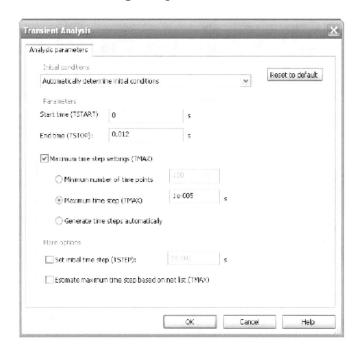

Figure 3.20: Specifications of the transient analysis.

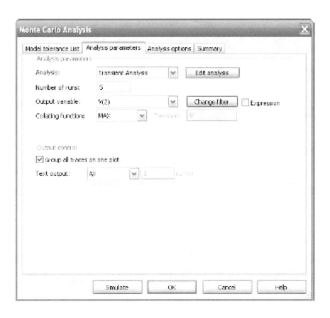

Figure 3.21: Specifications of the Monte Carlo analysis.

Feature	Description
MAX	Find the maximum value for each run.
MIN	Find the minimum value for each run.
RISE_EDGE	Find the first occurrence in that the response is greater than a specified threshold value in the rising edge of the waveform.
FALL_EDGE	Find the first occurrence when the answer is less than a specified threshold value in the falling edge of the waveform.

Table 3.1: Collating functions for Monte Carlo analysis.

From the spreadsheet data we can readily see that the Maximum Value deviation is given by

$$\text{Deviation} = \text{GreaterDeviationMaximum} - \text{NominalMaximum}$$
$$= 0.905441 - 0.862233$$

$$= 0.043208$$

The percentage nominal deviation is given by

$$\frac{\text{Deviation}}{\text{Nominal Maximum}} = \frac{0.047572}{0.862233} = 0.05011 = 5.01\%$$

This result agrees with the one given by Multisim in the last row of the spreadsheet in Figure 3.23. In another run with the same data, the statistics would be different due to the random nature of the Monte Carlo analysis.

Example 3.5 Active RC Chebyshev low pass filter using GICs.
Now let us consider the active RC Chebyshev low pass filter of Figure 3.24. This is an active realization of a low pass passive ladder prototype using generalized immitance converters (GIC) [1]. After an AC analysis, the frequency response of this circuit is shown in Figure 3.25a and the vertical axis is in linear scale and the horizontal axis in log scale. We can see there that the passband cut off frequency is 159 Hz. Figure 3.25b shows the 1 dB passband ripple (the frequency axis is in a linear scale and the vertical axis is now in a dB scale). We see here that the passband ripple has a 1 dB of variation.

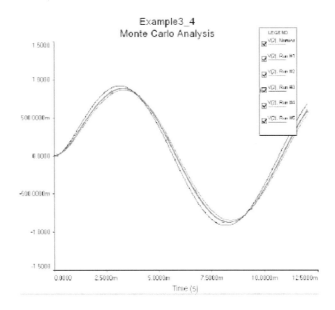

Figure 3.22: Plots of output waveform for all the runs.

Example3_4
Run Log Descriptions

	# of run	time [sec]	output value
1	Nominal Run (Mean Value for output : 0.866966 Standard Deviation for output: 0.0236087)	0.0033473	0.862233 (same as nominal, lower than mean by -0.00473 292)
2			
3	Run #1		0.833888 (3.28732% lower than nominal, lower than mean by -0.0330772)
4			
5	Run #2		0.854841 (0.8573% lower than nominal, lower than mean by -0.0121248)
6			
7	Run #3		0.889031 (3.10798% higher than nominal, higher than mean by 0.0220651)
8			
9	Run #4		0.85635 (0.68227% lower than nominal, lower than mean by -0.0106157)
10			
11	Run #5		0.905451 (5.01239% higher than nominal, higher than mean by 0.0384856)
12			

Figure 3.23: Plots of output waveform for all the runs.

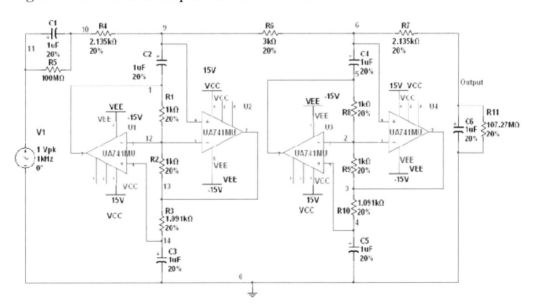

Figure 3.24: Active RC Chebyshev low pass filters using GICs.

We now perform a Monte Carlo analysis with all the passive components having a 20% tolerance with a Gaussian distribution (see

Figure 3.26). We only look at the passband responses. The output voltage is V(Output) and the number of runs is 5. This data is shown in the Analysis Parameters tab of Figure 3.27. After entering these parameters we click on the Simulate button to obtain the plots of Figure 3.28. The run at nominal values is thicker than the other runs. We see a deviation from the 1 dB ripple than in some runs is less than 1 dB and in other cases is almost 2 dB and in a case is even larger.

3.3.2 WORST CASE ANALYSIS

For a worst case analysis we need to carry out either a DC operating point or an AC analysis. In this analysis we are interested in finding out which is the run with the maximum deviation from the nominal run, either with an increasing or a decreasing deviation. We show the procedure with an example.

Example 3.6 Worst Case analysis for a passive RC filter.
We show the worst-case analysis procedure with the RC circuit shown in Figure 3.29. We select Simulate →Analyses →Worst Case which opens the dialog window in Figure 3.30. There we assign the tolerances.

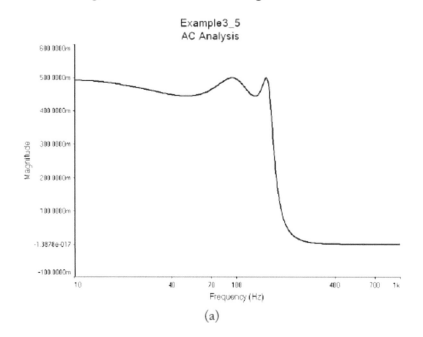

(a)

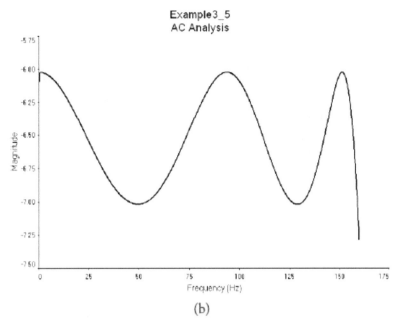

Figure 3.25: Frequency response for the active RC filter. a) Passband and stopband, b) Passband ripple.

Figure 3.26: Tolerance and distribution for the passive elements.

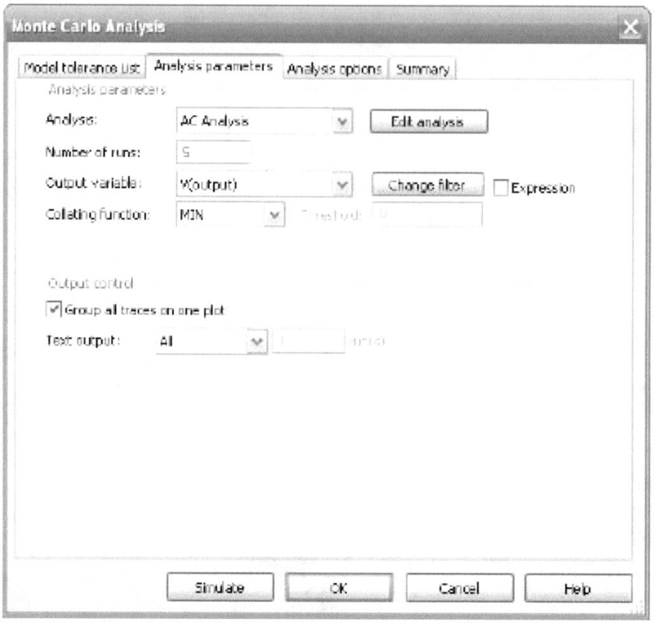

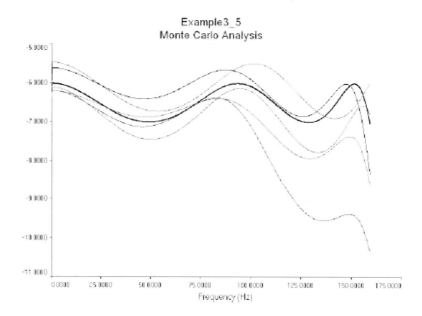

Figure 3.27: Parameters for the Monte Carlo analysis.

Figure 3.28: Passband ripple after the Monte Carlo analysis. Note that one of the runs produces almost a 6 dB variation in the passband ripple, originally set at 1 dB.

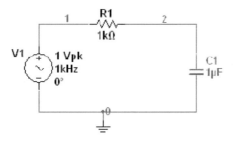

Figure 3.29: Passive RC filter.

After giving this data we pressed OK and we run Multisim. The output data Multisim gives us are the nominal value and the value of maximum variation from the nominal value. The results are shown in Figure 3.33. In Figure 3.34 we show the spreadsheet with the numerical results.

From the spreadsheet data we see that for the frequency response evaluated at 10 KHz, the magnitude has changed by 23.4486 % of the nominal value !!

Figure 3.30: Tolerance and distribution for the two components in the circuit.

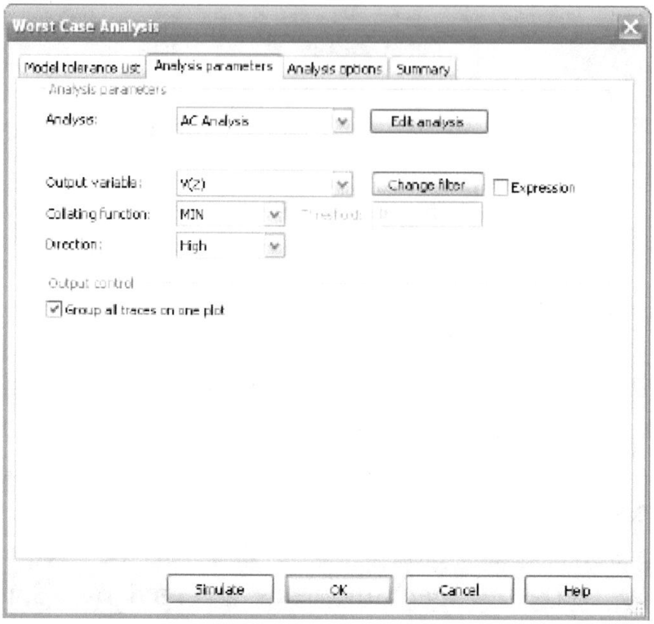

Figure 3.31: Data for Worst Case analysis.

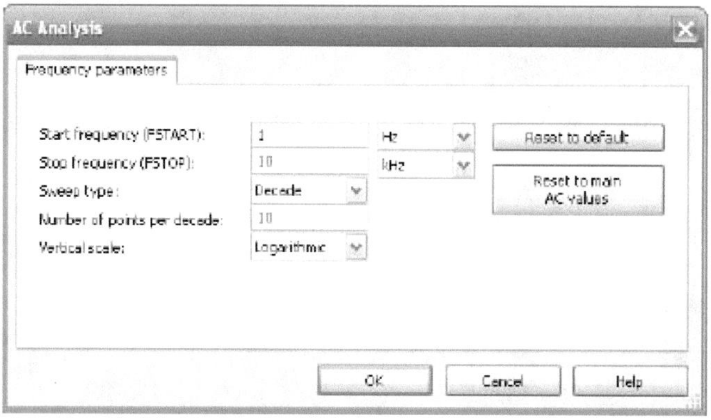

Figure 3.32: Parameters for AC analysis.

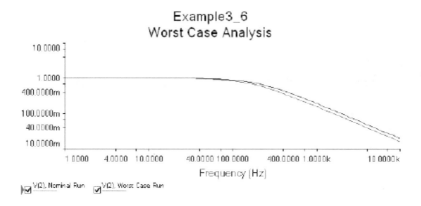

Figure 3.33: Plot from the worst case analysis.

3.4 PARAMETRIC ANALYSIS

A parametric analysis allows us to make any analysis changing the value of an element (parameter) within a specified range. For example, we might be interested in finding out the behavior of the circuit when a resistor changes value in a given range or to see what is the effect of changing β in a transistor amplifier. To run a parametric analysis, from the main menu select Simulate →Analyses →Parameter Sweep. This opens the dialog window of Figure 3.35 where we specify the parameter to sweep which can be either an element value (Device Parameter) or a parameter in a model (Model Parameter), for example, the β or the base emitter voltage of a transistor. We also specify the device type which can be a resistor, capacitor, etc., its name and its parameter (for example the β of a transistor, the nominal value of resistance, etc.). Also we specify the sweep type (Linear, decades, octaves, list), and the starting, final values, and increment of the parameter in the sweep and finally the analysis to be performed. We show the procedure with an example.

Run Log Descriptions

	Descriptions of the runs
1	Worst Case Run
2	AC analysis for all devices: 0.00373149 higher at frequency = 10000 (23.4486% of nominal)
3	
4	Tolerance changes needed to achieve worst case:
5	cc1 capacitance decreased to 9e-007
6	rr1 resistance decreased to 900

Figure 3.34: Spreadsheet with data from the worst case analysis.

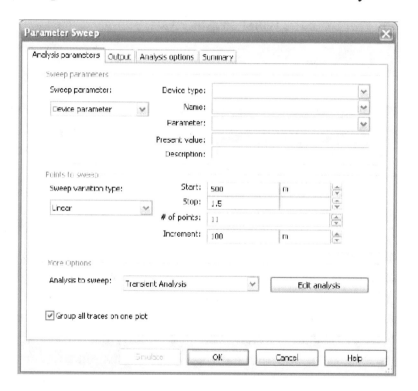

Figure 3.35: Dialog window to specify the parameters in a parametric analysis.

Example 3.7 Step response parametric analysis of an RLC passive circuit. Let us consider the RLC circuit of Figure 3.36. We wish to examine the transient response when the resistor changes value from its nominal value of 0.5 Ω. The range of resistor values is from 0.5Ω to 1.5Ω in increments of

0.1Ω. Select from the main menu Simulate →Analyses →Parameter Sweep and we enter the data as shown in Figure 3.37.

To specify the data for the transient analysis we click on the Edit Analysis button. We perform a transient analysis up to 14 sec as shown in Figure 3.38. We click on the OK button and we return to the window of Figure 3.37. In the Output tab we select the output variable as the current in the inductor (see Figure 3.39). We the click on the Simulate button and obtain the plots of Figure 3.40 where we see how the step response changes as the analysis is done for the different values of R change. We see that the step response changes from the nominal value where there is no overshoot to the response with R = 1.5Ω where the response shows a maximum overshoot.

3.5 TEMPERATURE EFFECTS

The temperature at which a circuit operates is important in its behavior. For this reason Multisim can also conducts analyses at different temperatures. All of the circuits we have simulated so far have parameters simulated at room temperature which is approximately 27° C. Devices that have a temperature dependency are resistors, inductors, capacitors, diodes, and transistors (BJT, JFET, and MOS). Multisim performs a temperature sweep changing the temperature at each step in the analysis. An example will show how a temperature analysis is made.

Example 3.8 Full Wave Rectifier.
A full wave rectifier circuit is shown in Figure 3.41. We make a temperature sweep to see how the response changes with temperature. Selecting from the main menu Simulate →Analyses →Temperature Sweep we obtain the dialog window of Figure 3.42. In this dialog window we specify the sweep analysis type which can be either Linear, by Octaves, by Decades, or by a List of temperatures. We specify a Linear analysis going from -25° C to 50° C with an increment of 5° C. Here we also specify the analysis type associated from the three basic analyses types with the button Edit Analysis. Since the voltage source frequency is 60 Hz, the final time in

the transient analysis TSTOP is 30msec where the temperature data for the analysis is entered (see Figure 3.43).

Figure 3.44 shows the ripple of the rectifier output. Here we see how the output ripple changes with temperature.

Example 3.9 Resistive voltage divider circuit.
Resistors have a linear and quadratic temperature dependence given by the following equation:

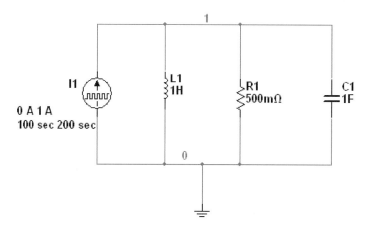

Figure 3.36: Circuit for parametric analysis.

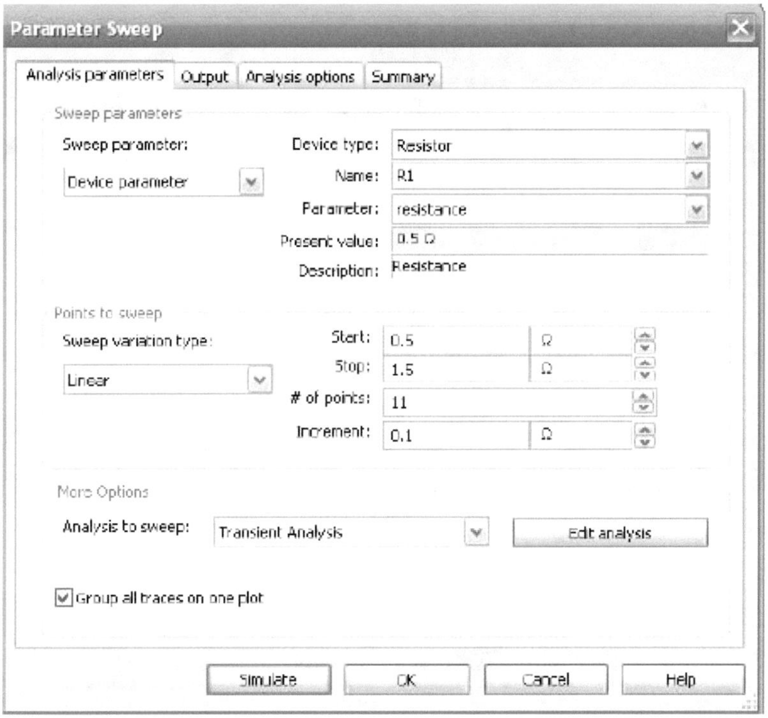

Figure 3.37: Dialog window with the values for the parametric analysis.

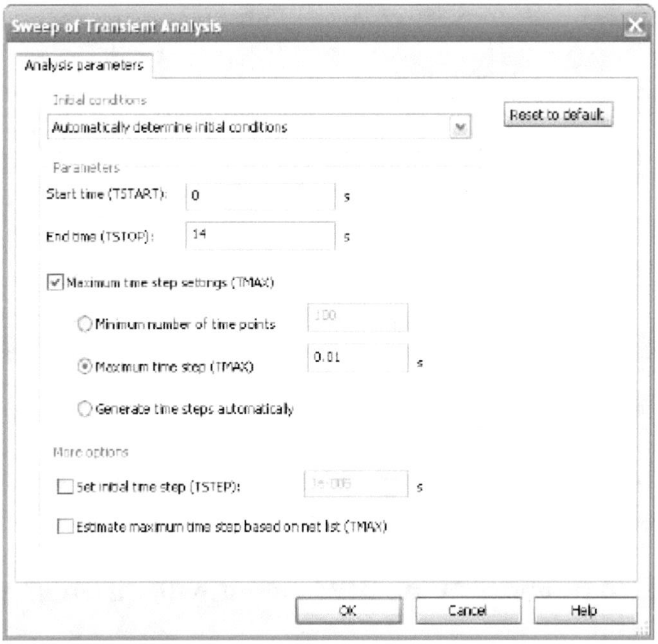

Figure 3.38: Data for the transient analysis.

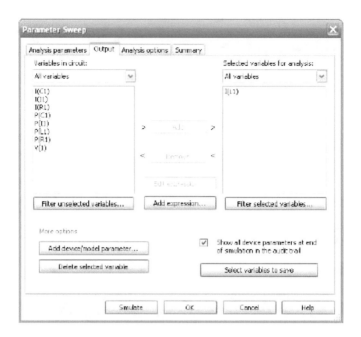

Figure 3.39: Specification of the output variable I(L1).

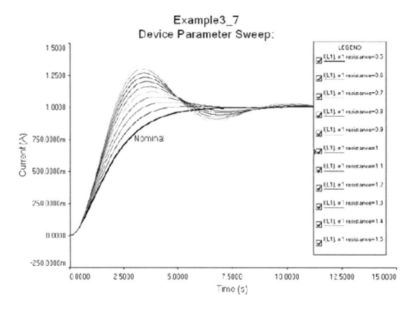

Figure 3.40: Step response of the RLC circuit with different resistor values.

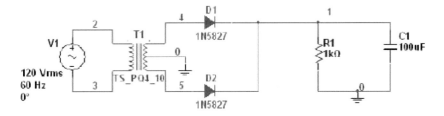

Figure 3.41: Full wave rectifier circuit.

$$\text{Resistor Value} = R\,[1 + \text{Resistor Value} = R\,[1 + TC1\,(T - Tnom) + TC2\,(T - Tnom)^2]$$

R is the resistor nominal value.

Tnom is the nominal temperature at 27°C.

T is the temperature at which analysis is done.

TC1 is the temperature linear coefficient.

TC2 is a quadratic temperature coefficient.

Usually, resistor manufactures specify linear temperature coefficients in the range of 100 to 1000 parts per million/°C (ppm/°C). This means that for a 1 kΩ resistor if the linear temperature coefficient is 1000 ppm then for each degree change the resistor value changes by

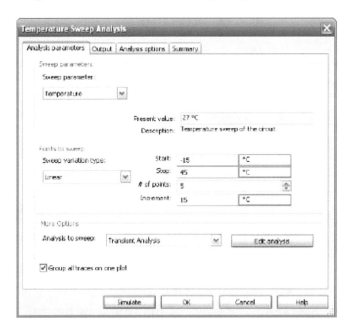

Figure 3.42: Temperature sweep dialog window.

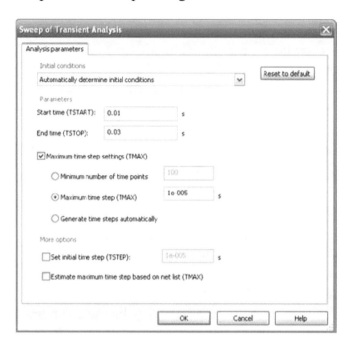

Figure 3.43: Data for the transient analysis.

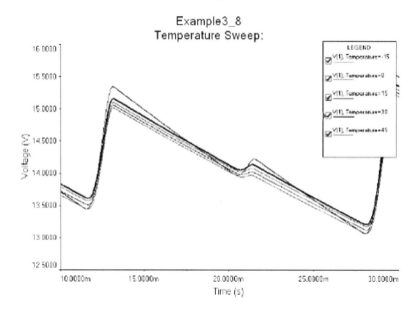

Figure 3.44: Temperature variation of the ripple in the full wave rectifier. Full wave rectifier circuit.

$$1000 \ \Omega \times 1000 \ \text{ppm/°C} = 1000\Omega \times 10^{-3}/°C = 1\Omega/°C$$

For TC1 = 100 ppm/°C, the change in resistor value for each degree change in temperature is 0.1 Ω/°C. Thus, the values of TC1 are usually very small. Similarly, the values for the quadratic temperature coefficient are very small. For Multisim, the value of TC1 is calculated as

$$coef \times ppm/°C = coef \times 10^{-6} /°C$$

where coef is the manufacturer specification for the temperature coefficient. Thus, if the manufacturer specifies a temperature coefficient of 200ppm/°C, the value of TC1 is 0.0002R Ω/°C, where R is the resistor value. Usually manufacturers only specify the linear temperature coefficient, but some carbon resistors present very small quadratic temperature dependence. Now let us consider the resistive voltage divider of Figure 3.45.

If we run a DC Operating Point Analysis, we obtain that V(2) = 6 volts. By clicking on resistor R1 we obtain the dialog window of Figure 3.46. In this window we see, besides the spaces for the resistor value and tolerance, the spaces for the the analysis temperature TEMP, the temperature coefficients TC1 and TC2, and the nominal temperature TNOM. For this example we specify the following values: TEMP = 50 °C, TC1 = 0.001, TC2 = 0.001, and TNOM = 27°C. After entering these values for R1 only we run again a DC Operating Point Analysis to obtain the value of voltage V(2) = 5.83885 as seen on Figure 3.47. This value is the result of the change in resistor value operating at a temperature of 50 °C. We can check the new value for R1 with the following:

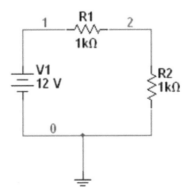

Figure 3.45: Voltage divider circuit.

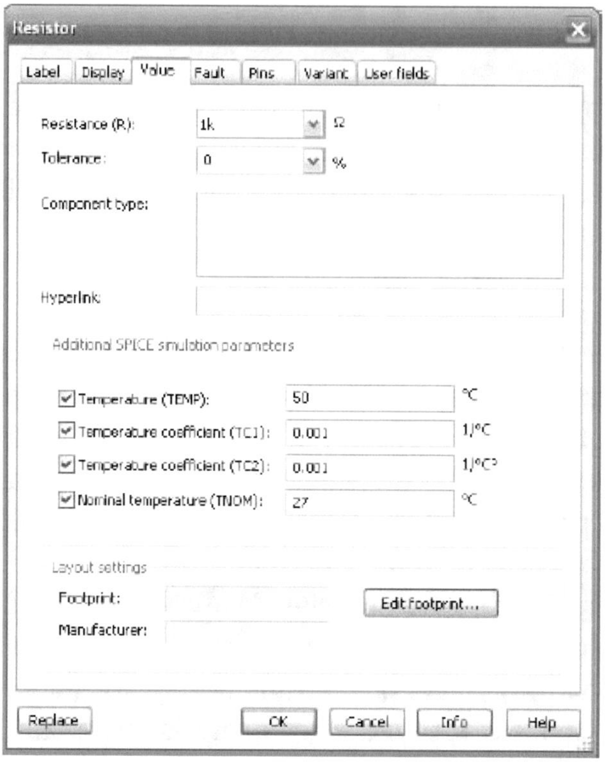

Figure 3.46: Dialog window to specify resistor temperature coefficients.

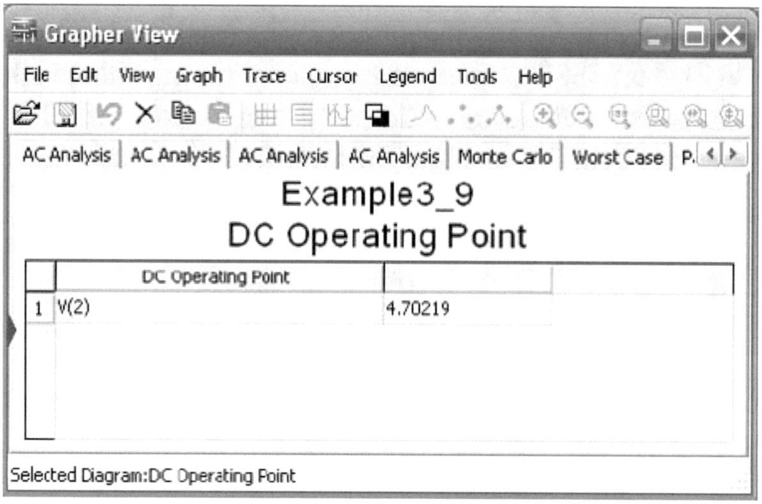

Figure 3.47: Voltage V(2) after running the analysis at 50°C.

Resistor value = 1000 [1+Resistor value = 1000 [1+ 0.001 (50 - 27) + 0.001
(50 - 27)^2] = 1552 Ω

Thus, the voltage divider value V(2) is

$$V(2) = 12\frac{1000}{1000 + 1552} = 4.702194$$

which is the same value as that obtained in the Multisim simulation.

3.6 CONCLUSIONS

In this chapter we covered three types of advanced analyses that complement the basic types of analyses. The analyses covered are the sensitivity analysis which is carried out jointly with the calculation operation point analysis (Bias Point), while the noise analysis is performed in conjunction with an AC analysis. The Monte Carlo analysis, worst case analysis, parametric and temperature analyses can be performed in conjunction with any analysis (AC, DC, transient).

3.7 PROBLEMS

3.1. For the circuit in Figure 3.48 perform a sensitivity analysis. The output voltage is the voltage across RL.

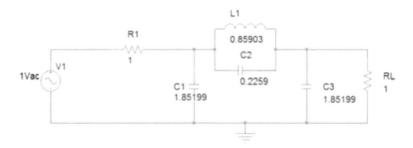

Figure 3.48: RLC circuit for sensitivity analysis.

3.2. In the circuit of Figure 3.49 make a Monte Carlo analysis after an AC analysis. Use 10% tolerances.

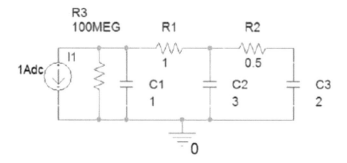

Figure 3.49: RC circuit for Monte Carlo analysis.

3.3. Repeat Exercise 3.2 but now make a Worst case analysis. Use 5% tolerance for the resistors and 10% for the capacitors.

3.4. The circuit of Figure 3.50 is an IF filter. Make a noise analysis and obtain the noise figure.

3.5. Make a temperature analysis for the circuit in Figure 3.51. Use : a) TC1 = TC2 = 0.001, b) TC1 = TC2 = 0.002.

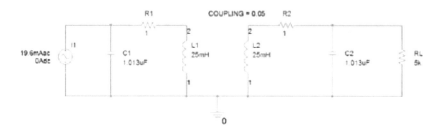

Figure 3.50: IF filter for worst case analysis.

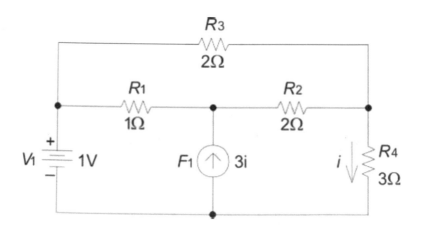

Figure 3.51: Circuit for temperature analysis.

CHAPTER 4

Simulating Microcontrollers

4.1 INTRODUCTION TO MULTISIM MCU

Multisim MCU is an add-on to Multisim that enables the user to build and simulate circuits based on microcontroller units. The letters MCU stand for Microcontroller Unit Co-Simulation for SPICE-Based Circuits

Microcontroller circuits are capable of handling several tasks (reading/writing data to a memory, interfacing sensors, acquiring real world signals, etc.) based on a program written by the user. Multisim MCU can simulate the behavior of a microcontroller running custom code.

The advantages of programming and testing in a simulated environment are numerous. One of the main benefits of this approach is the ability to run and debug microcontroller circuits *interacting* with electronic components. This is something that most debuggers fail to deliver.

4.2 MULTISIM MCU BASICS

In Multisim the procedure for working with microcontrollers is slightly different to the procedure for dealing with electric or electronic circuits directly.

The first requisite to be covered is to have already installed the MCU module. This module is part of the student version and it is responsible of providing the appropriate environment for microcontroller simulation.

In the physical world a microcontroller is usually programmed using dedicated hardware. Sometimes this hardware is programmed using an RS232 or USB programmer. Nowadays, even Bluetooth programmers can be used. The main function of the programmer is to transfer a .hex file (the .hex extension comes from *hexadecimal* because the source file is in binary/hexadecimal format) to the microcontroller. In order to do this, some considerations apply like holding the reset line at a low level and use the appropriate signal level (commonly TTL levels used are 0 and 5 VDC).

This binary/hexadecimal file is obtained after having compiled and linked the source code. This task is accomplished by the *compiler*. The compiler is a computer program that translates instructions in high/low level to the binary/hexadecimal format that can be electronically transferred to the microcontroller.

The compiler can receive instructions in several ways. The most common ways are using assembly language (low level language) or C language (high level language). For some microcontrollers the use of high level languages (like C or Basic) can have a negative impact on the general performance of the microcontroller. This impact is directly related to the fact that a larger translation takes place when using high level languages. Assembly code is very close in length to what the microcontroller receives when it is programmed. Some microcontrollers are designed from the beginning to have their architecture optimized when programmed with code made from higher level languages. They can have almost no performance hit from a practical standpoint.

In the case of Multisim, the built-in compiler is PICC-Lite from the company Hi-Tech. By modifying some options during the initialization of each project the programmer can choose to use the compiler of his/her choice.

4.2.1 INCLUDED MICROCONTROLLER MODELS

Included models in Multisim MCU are basically two – microcontrollers from the 8051 family and microcontrollers from the PIC16F84 family.

The microcontrollers in the former family were used extensively some years ago for embedded systems. Specifically, the 80C51 from Intel was designed with CMOS technology and consequently their energy consumption was lower than previous designs, making them more attractive for battery-operated systems.

These microcontrollers offered, in a single package, a device with some peripherals built-in: RAM, ROM, inputs, outputs, timers, interrupts and serial communication via UART or USART ports were common. Nowadays, several manufacturers offer more sophisticated products and, in some cases, they continue to produce updated versions of the 8051.

The other family of microcontrollers included in Multisim MCU is PIC16F84 and PIC16F84a from Microchip. These microcontrollers are more recent than the 8051 and they are designed with an 8-bit RISC architecture. The family 16X84 includes the 16C84 which is a microcontroller that stores its program in EEPROM memory. The 16F84 uses Flash memory for program storage.

In this chapter we are going to show examples using the PIC16F84. We will start using assembly language for the first example and C for the rest of the examples.

These microcontrollers can be connected to all of the electrical components found on the Parts bin of Multisim. They can also be connected to some advanced peripherals found in the Advanced Peripherals section of the Place menu. There are different types of RAM and ROM memory, numeric keypads for data input, liquid crystal displays for data output and some other parts we are going to use later on.

4.2.2 MCU WIZARD

The MCU Wizard assists the user in the creation of projects that involve microcontrollers. It will start the moment you add a microcontroller from the Place Part menu. The procedure is summarized as follows:

1. Place → Component → MCU. Here we choose the PIC type.

2. The MCU Wizard starts

3. The workspace path can be anywhere in your hard drive. We suggest using the folder where we are storing our examples, for example *My Documents → Multisim* (Figure 4.1).

4. The workspace name for our first example is named accordingly, *example01* (Figure 4.1).

5. The next step (2 of 3) defines the project settings. This time we will use a standard project type. The programming language should be adequate for the language we plan to use, and so it should be the compiler and the project name can be anything meaningful, for example *Blinking LED* (Figure 4.2).

6. The last step (3 of 3) allows you to start with an empty source file or with a predefined source file. We will use main.asm (or main.c) as our source file (Figure 4.3).

Figure 4.1: MCU Wizard.

Example 4.1 Blinking an LED (using assembly language).
When learning how microcontrollers are programmed, turning an LED on and off is the equivalent to programming a simple *Hello World* program when learning a programming language. This time we will do the same as our first example.

We start by opening a new circuit and placing a PIC16F84 microcontroller from the Place Parts menu. The steps have already been outlined above, but we detail a little bit more here.

Figure 4.2: MCU Wizard step 2.

Figure 4.3: MCU Wizard step 3.

The suggested values for our first project can be seen in Figures 4.1 to 4.3. This is the complete procedure for configuring a microcontroller using the MCU Wizard.

Now we select the type of project. This is a standard project. The other option allows us to use a pre-compiled hex file that Multisim do not need to recompile again. We do not use this feature here. We select assembly language as our programming language and the compiler automatically shows Microchip MPASM for PIC 16. An appropriate name for the project is *Blinking LED*.

Figure 4.4: Design Toolbox showing the tree view of a MCU project.

In the next window we are asked if we want to create a blank project or add a source code file (Figure 4.3). For our needs, starting with a source code file named *main.asm* is perfectly acceptable.

Up to this point we can see the tree view for the project in the Design Toolbox window. You can access it via View → Design Toolbox (Figure 4.4). Inside this window we can see the structure of our project and add or delete additional files.

If we double click on each branch of the tree view we can display the content of the file *main.asm* and start to edit the code. The code is exactly the same as if we were programming a physical microcontroller. In this particular case, a table of commands can be found on the manufacturer's site:

`http://ww1.microchip.com/downloads/en/DeviceDoc/33023A.pdf` and in the microcontroller's datasheet

`http://ww1.microchip.com/downloads/en/DeviceDoc/35007b.pdf`

We use the following code:

```
#include ''p16f84.inc''   ; This includes PIC16F84 definitions for
the
                          ; MPASM assembler
;main routine
```

```
;enable PORTB as output

    CLRF PORTB              ; Initialize PORTB by
    ; clearing output
    ; data latches
    BSF STATUS, RP0         ; Select Bank 1
    MOVLW 0x00 ; Value used to
    ; initialize data
    ; direction
    MOVWF TRISB             ; all port pins are outputs
    BCF STATUS, RP0         ; Back to Bank 0

main
    MOVLW 0x01
    MOVWF PORTB
     nop
     nop
     nop
     nop
    CLRF PORTB
     nop
     nop
     nop
     noP
    GOTO main
    END
```

Once we have typed or copied the code to the *main.asm* file we can build it (in the menu → MCU → MCU PIC16F84 U1 → Build. In the spreadsheet window we can see the error log and warnings, if any. In Figure 4.5 we can see the tab for Results.

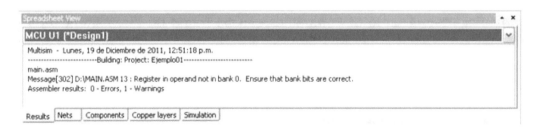

Figure 4.5: Spreadsheet view.

Now that our microcontroller has been programmed we can make the electrical connections. This time we will use a VDD power source and a VSS ground reference. We will also use a digital indicator connected to pin RB0. Note that the Master Clear pin is inverted so it must be tied to VDD for our circuit to operate properly (Figure 4.6). Now we run the circuit by clicking on the run icon (or pressing F5).

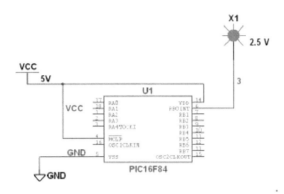

Figure 4.6: Electrical connections.

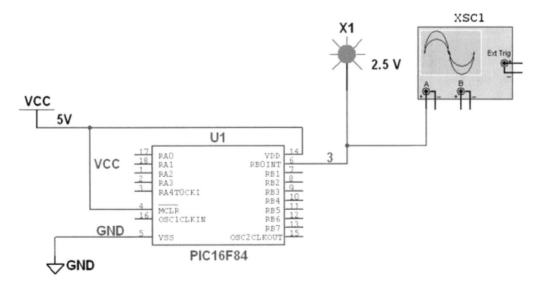

Figure 4.7: Blinking LED.

Notice how the circuit does not have an oscillator connected to its respective pins. The inevitable question is how can we verify/modify the frequency of oscillation for this circuit? The answer can be found right clicking on the microcontroller. In the Value tab we can see that the clock speed is set to 12 MHz. We can make adjustments if necessary but we will leave it as it is for the moment (Figure 4.8).

Figure 4.8: Setting the clock speed.

If we have correctly programmed the microcontroller and made the right electrical connections now it is time to run our simulation and see if the LED blinks (Figure 4.7).

As you should have noticed we added an oscilloscope to the circuit. Probably you are seeing the LED flashing very quickly and we would like to measure the flickering frequency. Using cursors on the scope's window we can see that the signal's period is 4.32 us thus the frequency is approximately 231 KHz. We can slow it down modifying the code. Try including more *nop* commands. They are used when you want to waste time. Basically, a nop command tells the microprocessor to execute *no operation.* The same measurement can be done using the frequency meter from the instruments toolbar.

Example 4.2 Blinking an LED (using C language).
Our first example showed us how to set up a project for assembly language programming. This time we will do the same but with the difference of using C as our programming language. The advantage of using C instead of assembly is that due to it being a higher-level language it's easier to understand and follow up when reading the code. The disadvantage is that C needs to be translated to assembly by the compiler and this translation sometimes may not be the most efficient way of doing it. In the next section we will see how we can use the debugging tools for optimizing the generated code.

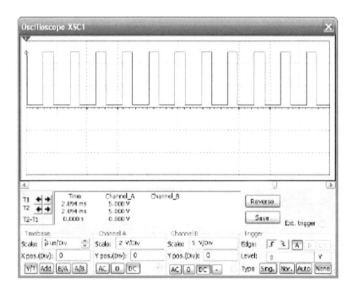

Figure 4.9: Oscilloscope with blinking signal to measure the frequency.

This time we will use the following code:

```
#include "pic.h"

#define bitset(var,bitno)     ((var) |= 1 << (bitno))
#define bitclr(var,bitno)     ((var) &= ~(1 << (bitno))

// a delay function for slowing down the execution
```

```
void delay(void)
{
  int count = 0;
  for (count = 0; count < 250; count++) {          }
}

void main()
{
    bitclr(STATUS, RP0); // Select Bank 0
    bitset(STATUS, RP0); // Select Bank 1
    TRISB = 0x00;        // Set Port B pins to output mode
    while(1)
    {
      PORTB=0x01; // Turn on RB0
      delay();    // Wait
      PORTB=0x00; // Turn off RB0
      delay();    // Wait again
    }
}
```

Figure 4.10: Design Toolbox for a C-language project.

We can see that the code is much more compact this time. We now should build the code and run it. Whenever the user wants to run an out-of-date program, Multisim will prompt the user to re-build it. When executed we can see that the LED flashes more slowly this time. In this case, the oscillation frequency can be adjusted by modifying the value of the counter in the `delay()` function. With different values we can make the LED flash at different speeds. The variable *count* is an integer and thus it can contain values from -32767 to 32766. If we try to count beyond its limits we will see that it rolls up. This is something that programmers need to be careful.

Going into further details of assembly/C programming is out of the scope of this book. An excellent reference for going deeper into C is found on the book *Practical C Programming*, by Steve Oualline, O'Reilly 1991 [1].

4.3 DEBUGGING TOOLS

Multisim has several useful debugging tools when using the MCU module. These tools help the developer to test his/her program and find errors that frequently are hard to detect and fix when the microcontroller is freely running.

In the menu MCU → MCU Windows we can activate the Memory View window. In this window we can see the memory content of the microcontroller. For Example 4.2 it is shown in Figure 4.11. We see the value of each one of its registers and the configuration data. This memory view allows the programmer to inspect the contents of the memory when verifying the behavior of a particular program.

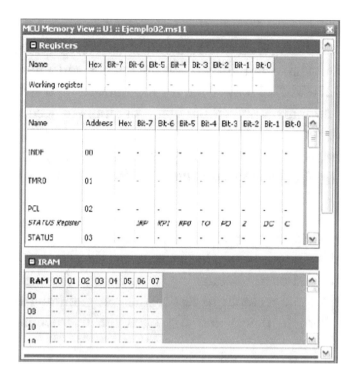

Figure 4.11: Memory contents.

Another frequently used task when debugging programs is stopping it from continuous execution. Being able to stop the program in an arbitrary step of the code allows the programmer to inspect the state of the memory and the related variables. This way a particular condition can be verified and errors can be corrected.

The MCU module has several variants of this function included in the *Debug View* menu. This function can be activated going to MCU → name of the microcontroller (PIC16F84 U1 in these examples) → Debug View (Figure 4.12).

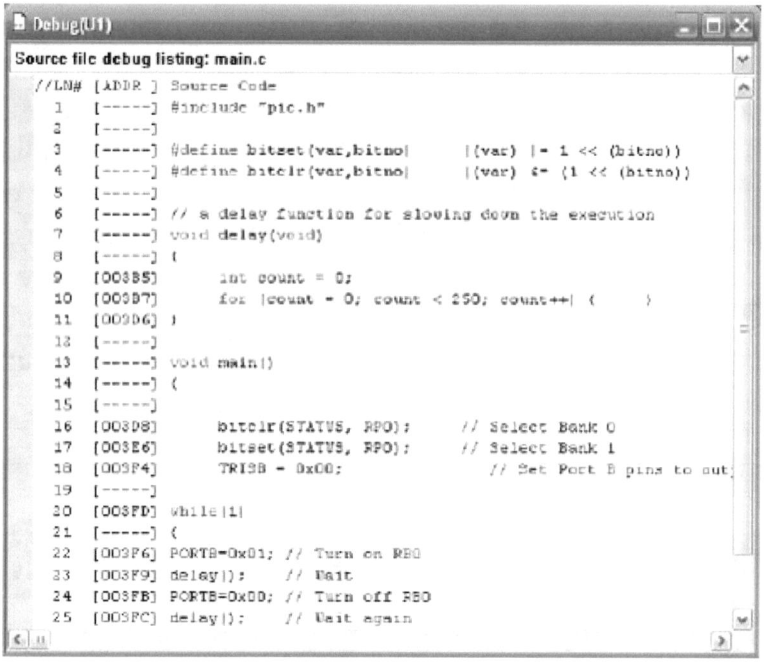

```
Debug(U1)                                                    _ □ X
Source file debug listing: main.c                              ✓
  //LN# [ADDR ] Source Code                                    ▲
    1   [-----] #include "pic.h"
    2   [-----]
    3   [-----] #define bitset(var,bitno)    |(var) |= 1 << (bitno))
    4   [-----] #define bitclr(var,bitno)    |(var) &= (1 << (bitno))
    5   [-----]
    6   [-----] // a delay function for slowing down the execution
    7   [-----] void delay(void)
    8   [-----] {
    9   [003B5]     int count = 0;
   10   [003D7]     for (count = 0; count < 250; count++) (      )
   11   [003D6] }
   12   [-----]
   13   [-----] void main()
   14   [-----] {
   15   [-----]
   16   [003D8]     bitclr(STATUS, RP0);    // Select Bank 0
   17   [003E6]     bitset(STATUS, RP0);    // Select Bank 1
   18   [003F4]     TRISB = 0x00;           // Set Port B pins to out
   19   [-----]
   20   [003FD] while(1)
   21   [-----] {
   22   [003F6] PORTB=0x01; // Turn on RB0
   23   [003F9] delay();    // Wait
   24   [003FB] PORTB=0x00; // Turn off RB0
   25   [003FC] delay();    // Wait again                        ▼
```

Figure 4.12: – Debug View window showing source file code.

In the Debug View mode we can observe the disassembled code (Figure 4.12). If we are using assembly language for programming this view is very similar, the main difference is that each variable is substituted by its real value. *Goto* and *Jump* instructions keep their labels.

When programming using C the disassembled code will show us how the compiler is interpreting the C code and how it is translated to assembly code. This is a great way to learn about going from a higher level language to a low level language like assembly.

In the top part of the Debug View window we can select the option if we want to debug using disassembled code or the source code. The main difference is that the source code contains labels and comments. Sometimes is more useful to work with fully-documented code.

When the debug mode is active we can stop the program when it executes each instruction. This is accomplished using the *Step Into* function; this function is conveniently mapped to the F11 key. F10 key is mapped to the *Step Over* function. When we use these two functions we can execute the program one step at a time and use Memory View to observe what is happening with the values of different registers and memory locations.

Similarly, when in debug mode, if we have activated the MCU toolbar (View → Toolbars → MCU, and shown in Figure 4.13) we can select some relevant options that display the code in specific ways. A listing of the tasks that can be done is given in Table 4.1 C code when disassembling so we can verify the correspondence. We can enable/disable the memory addresses. We can see the hexadecimal values for each *opcode* – opcodes are the numeric equivalent of each one of the commands of the instruction set of the microcontroller. We can show labels and headings as well.

Table 4.1: Buttons in the Debug toolbar.	
Button	**Description**
DASM	**Disassembly** button. Determines whether we see the listing assembly text or the version from the disassembler.
	Show Secondary Language as Comments button. Enables/disables one of two things. If we are looking at the project disassembly view or if we are looking at the listing assembly or disassembly of a file it adds the original source code corresponding to those lines as comments above or beside the assembly. If we are viewing the source code debug listing, this shows the corresponding listing assembly/disassembly as comments below the source code.
	Show Line Numbers button. Displays the line numbers in the **Debug View**.
	Show Memory Addresses in Debug View button. Enables/disables the showing of memory addresses for any text that shows source code in any of the debug listing views.
	Show Memory Addresses in Assembly Code button. Enables/disables the showing of memory addresses for any text that shows listing assembly or disassembly in any of the debug listing views.

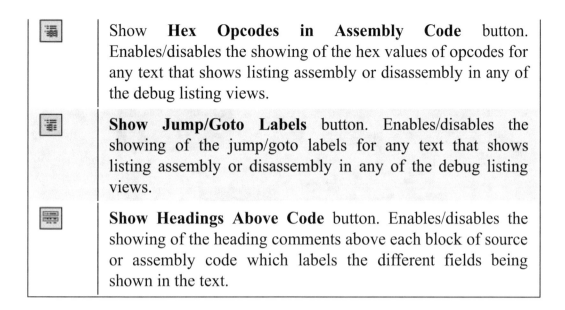	Show **Hex Opcodes in Assembly Code** button. Enables/disables the showing of the hex values of opcodes for any text that shows listing assembly or disassembly in any of the debug listing views.
	Show Jump/Goto Labels button. Enables/disables the showing of the jump/goto labels for any text that shows listing assembly or disassembly in any of the debug listing views.
	Show Headings Above Code button. Enables/disables the showing of the heading comments above each block of source or assembly code which labels the different fields being shown in the text.

4.4 PERIPHERAL DEVICES

Multisim MCU Module offers several peripheral devices for using along with the included microcontroller models. Among them we can find RAM memory, ROM memory, liquid crystal displays, numeric keypads, a virtual serial terminal, a conveyor belt, a holding tank and two different traffic lights (Figure 4.14).

Figure 4.13: MCU toolbar.

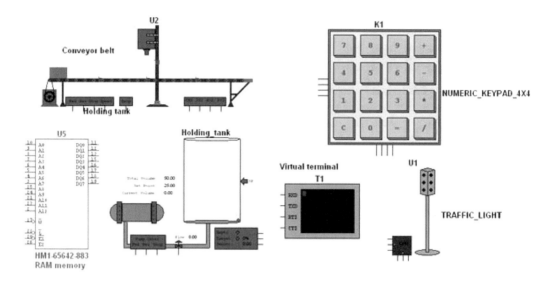

Figure 4.14: Example of the included peripherals.

4.5 EXAMPLES

In this section we will show three examples using the elements we have discussed so far. We will be using C for these examples but the reader should feel free to adapt them to use assembly language if desired.

The set of experiments will include this:

- Even/Odd counter

- LCD Display

- Traffic Light

Example 4.3 Even/Odd counter.

In this example, a microcontroller will receive user input on a pin by the way of an electromechanical switch and the result will affect a count shown on a HEX display.

The circuit is shown on Figure 4.15.

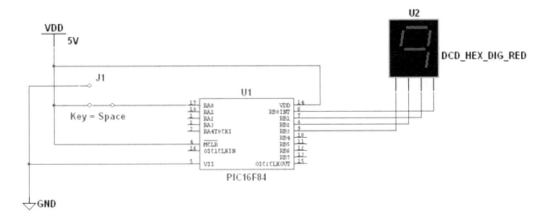

Figure 4.15: Even/Odd counter.

As you can see from the previous figure we have included two elements in this circuit –the switch and the display. The switch can be found in the Basic Components → Switches bin. This switch in particular is a SPDT switch – single pole, double throw. Similarly, the display can be found in the Indicators → Hex Display bin. The type, as is shown on Figure 4.15, is DCD_HEX_DIG_RED.

The switch, as the label shows, is controlled by the space key, this basically means that whenever we press the space key the switch will change from closed to open back and forth. The display receives 4 bits as input and decodes them as a BCD quantity and shows the appropriate symbol.

In order to accomplish this task we need to have a program with the following capabilities:

- Read from the input pin RA0 and write a count to Port B. This configuration is made by setting the right bits to the TRISA and TRISB registers.

- Count from 0 to 15

- Modify the count to show only even numbers or odd numbers depending on the switch's position

The proposed solution is a program that addresses directly these requirements. First we will configure the microcontroller for input on Port A and output on Port B. Then we will use a For loop for counting from 0 to 15 (that is, from 0x0 to 0xF in HEX notation) and we will write data to the output port using the input on Port A pin 0 (RA0) as a modifier so it can show even numbers or odd numbers. There is also an ancillary function called delay so the count does not flash too quickly on the display.

The proposed solution is shown next:

```
#include "pic.h"

#define bitset(var,bitno)    ((var) |= 1 << (bitno))
#define bitclr(var,bitno)    ((var) &= ~(1 << (bitno))

int i;

// delay function to slow things down
void delay(void)
{
  int count = 0;
  for (count = 0; count < 250; count++) {      }
}

void main()
{
  bitclr(STATUS, RP0); // Select Bank 0
  PORTB = 0x00;        // Clear Port B
  bitset(STATUS, RP0); // Select Bank 1
  TRISB = 0x00;        // Set Port B pins to output mode
  TRISA = 0xFF;        // Set Port A as inputs

  while(1)
  {
    for(i=0;i<16;i+=2)
    {
      PORTB=i-RA0; // We subtract the value of RA0 so we can
      delay();     // make an even or odd count
    }
  }
}
```

When this program is built (compiled and linked) we can start the simulation and see how the display counts odd numbers when the switch is connected to 5V and when it is tied to ground it will count even numbers. The book's website shows an animation of this circuit.

Example 4.4 Writing data on an LCD display.

The following example will make use of the liquid crystal display (LCD) included on the Advanced Peripherals library. This kind of displays is based on a very common controller –the Hitachi 44780. Manufacturers still use most of the original specifications so they are interoperable and backwards compatibility is not an issue.

Figure 4.16: LCD terminals.

As we can see in Figure 4.16 there are several connection pins on this module. We have the usual supply pins (VCC and GND), a control voltage (CV) used for adjusting the contrast, 3 control pins (E, RS and RW) and 8 data pins (D0 . . . D7). We will connect them as follows:

- VCC at 5VDC

- CV, GND and RW at 0VDC

- E and RS at RA0 and RA1, respectively,

- D0 to D7 directly connected to Port B

In order to display characters on an LCD we must first initialize it. Our program will execute the initialization routine upon start. Then we will define an ASCII string that will be sent and displayed.

The complete circuit is shown on Figure 4.17. As you can see we have used LED indicators as a way to visually inspect that the digital signals are being sent as expected. Also now we are using a more common display type, a 16x2 characters display. That means it has two rows, with 16 characters each one.

The program for the PIC is:

```
#include "pic.h"

#define bitset(var,bitno)     ((var) |= 1 << (bitno))
#define bitclr(var,bitno)     ((var) &= ~(1 << (bitno))

void delay(void)
{
  int count = 0;
  for (count = 0; count < 250; count++){     }
}

// toggle function toggles the value at RA0

void toggle(void)
{
  RA0=0;
  RA0=1;
}

// lcdInit function is the initialization procedure for a 16x2 display
void lcdInit(void)
{
  RA1=0;
  delay();
  PORTB=0x38; // configures the display as 16x2, 5x7 font and 8 bit
 bus
  toggle();
  PORTB=0x06; // moves cursor to the right
  toggle();
  PORTB=0x0F; // turns on the display
  toggle();
  PORTB=0x01; // clears and sets the cursor at the home position
  toggle();
  delay();
  RA1=1;
}

void main()
{
  bitclr(STATUS, RP0); // Select Bank 0
  bitset(STATUS, RP0); // Select Bank 1
  TRISA = 0x00;        // Set Port A pins to output mode
```

```
    TRISB = 0x00;           // Set Port B pins to output mode
    lcdInit();
    delay();
    PORTB = 0x45; // writes letter E
    toggle();
    PORTB = 0x41; // writes letter A
    toggle();
    PORTB = 0x53; // writes letter S
    toggle();
    PORTB = 0x59; // writes letter Y
    toggle();
    PORTB = 0x20; // writes a space
    toggle();
    PORTB = 0x41; // writes a letter A
    toggle();
    PORTB = 0x53; // writes a letter S
    toggle();
    PORTB = 0x20; // writes a space
    toggle();
    PORTB = 0x41; // writes a letter A
    toggle();
    PORTB = 0x42; // writes a letter B
    toggle();
    PORTB = 0x43; // writes a letter C
    toggle();
    PORTB = 0x20; // writes a space
    toggle();
}
```

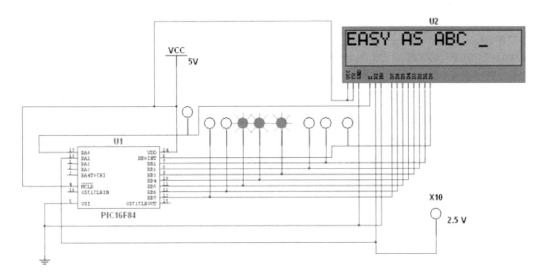

Figure 4.17: Circuit for testing an LCD.

We can see a pair of functions at the top of the program: `toggle()` and `lcdInit()`. The toggle function just sends a pulse going from low to high whenever we want a new data to be fed into the display. The delay function is the same we have used in the previous Examples 4.2 and 4.3.

When the program is compiled and the simulation starts we can see a quick flash of the LEDs and after a while we can see that letters start to appear on the display. Since this is a small message we are storing it right in the microcontroller's memory. In a real world situation messages are usually longer and need to be stored in an external memory and read as needed.

The data in the main section of the program is the message we want to display. In this case, the message reads *EASY AS ABC*. Each character is coded into its ASCII representation in hexadecimal notation. The used sequence (in hexadecimal) is:

45-41-53-59-20-41-53-20-41-42-43

As a practical exercise try to find out the code for the space, for the letter A and, using an ASCII chart, try to see how to convert this string to all lowercase characters (*Hint: they are separated a fixed distance*).

Now we will focus our attention on the `lcdInit()` routine. The purpose of this function is to prepare the LCD for the right number of rows and characters, clear the screen and position the cursor where we want. In Multisim we can get along without initializing the display just fine, but anyway it is a good practice –and certainly a requirement most of the time – to always initialize a display before using it.

If we look carefully at the connections in Figure 4.17 we see that there is a control pin labeled RW. This signal is always tied to ground because we will be writing *to* the display. The reason for it to be there is that a read/write procedure is used when the user wants to define custom characters or reading the status of the several function of the display during execution.

Example 4.5 Controlling a traffic light.

The next example will make use of the included traffic light from the Advanced Peripherals/Misc Peripherals menu. This device along with the conveyor belt and the holding tank are part of the educational edition of

Multisim and are mostly used when teaching and learning about ladder logic. As a matter of fact, microcontrollers and PLCs (programmable logic controllers) are commonly used when working in similar situations in the real world so what you learn in this chapter can also be applied to a broad range of topics.

A microcontroller can only output very small currents. It is common to see maximum ratings in the order of 40 mA on each pin. This current is not enough for powering a lot of common devices, like the lights on a traffic light. We must find another way of controlling a large current with a small control signal. In this specific case we are going to use a component called *voltage controlled switch*. This component can be found under the Components → Basic → Switch menu as we see on Figure 4.18.

Figure 4.18: Voltage controlled Switch.

This switch is special because it gives us the power to control devices that need a current larger than a few miliamperes. Most of the time these switches are made of transistors or electromechanical relays. We will use it as the switching element for controlling the inputs of the Traffic Light that can be found on the Advanced Peripherals menu. This traffic light turns on each one of its lights whenever a voltage is present at the input. If we would have chosen the microcontroller pins as the control signals we would have seen that the lights never turn on, or perhaps they flash very quickly. A real microcontroller would not withstand such operation and a resulting damaged pin would be the consequence of this operation.

The connections are shown on Figure 4.19. The control signals come from the microcontroller and they are used as inputs on the voltage controlled switches. These switches have their threshold voltage modified to turn on and off at 2.5 volts.

We can also see that the other side of the switch has one connection going into the traffic light and the other side is left open.

The microcontroller is programmed with a very basic main function. This does not mean it is a simple function. Here you can see it:

```
void main()
{
  bitclr(STATUS, RP0); // Select Bank 0
  bitset(STATUS, RP0); // Select Bank 1
  TRISB = 0x00;        // Set Port B pins to output mode
  state01();           // red on one side and green on the other
  blinkGreen01();      // green light 01 blinks
  amber01();           // still red on one side and amber on the ot
her
  state02();           // this state is like state 01 but inverted
  blinkGreen02();      // green light 02 blinks
  amber02();           // amber light on one side and red on the ot
her
}
```

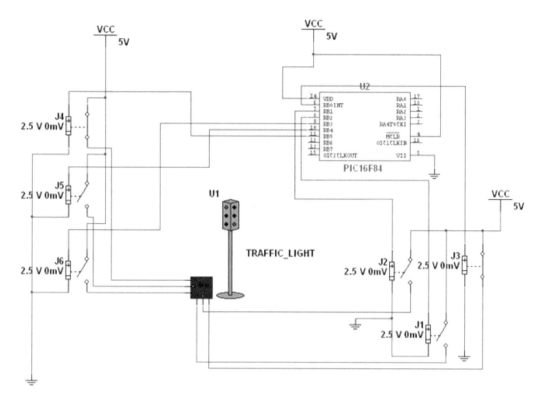

Figure 4.19: Circuit for a traffic light controller.

The first part is just the setup of the PORTB. Then we see a sequence of functions *state01, blinkGreen01, amber01, state02, blinkGreen02,* and *amber02*. These functions are defined in the previous part of the program and it is there where the action happens. The main function was intentionally left as just a way of calling the functions just to explain more clearly the sequenced nature of this problem. We could have written every step as part of the main function and the program would have run exactly the same.

The operation of the traffic light is defined as follows:

1. A green light on one side is accompanied by a red light on the other side.

2. After some seconds, the green light starts to blink briefly before going into amber.

3. When the amber light turns off, the red light and the green light are reversed with respect to the initial state.

4. The blinking and amber lights happen on one side as the other holds the red light on.

5. Everything repeats again.

The complete program is shown next. There are three different delay functions. These may need tuning when simulating this example on different computers since their timing depends on the processor's speed.

```
#include "pic.h"

#define bitset(var,bitno)      ((var) |= 1 << (bitno))
#define bitclr(var,bitno)      ((var) &= ~(1 << (bitno))

// this is a standard delay function
void delay(void)
{
  int count;
  for (count=0;count < 255;count++) {      }
}

// this delay is a little bit longer than the previous one
void delayMedium(void)
{
  int count;
  int count2;
  for (count=0;count < 255;count++) {
    for (count2=0;count2 < 3;count2++) {      }
  }
}

// this is the longest delay
void delayLong(void)
{
  int count;
  int count2;
  for (count=0;count < 255;count++)
  {
    for (count2=0;count2 < 10;count2++) {      }
  }
}
```

```c
void state01(void)
{
  PORTB=33;      // red on one side and green on the other
  delayLong(); // long delay time on this state
}

// Green light 01 blinks four times
void blinkGreen01(void)
{
  PORTB-=1;
  delay();
  PORTB+=1;
  delay();
  PORTB-=1;
  delay();
  PORTB+=1;
  delay();
  PORTB-=1;
  delay();
  PORTB+=1;
  delay();
}

// amber light on and red on
void amber01(void)
{
  PORTB=34;
  delayMedium();
}

// like state01 but inverted
void state02(void)
{
  PORTB=12;
  delayLong();
}

// green light 02 blinks four times
void blinkGreen02(void)
{
  PORTB-=8;
  delay();
  PORTB+=8;
  delay();
  PORTB-=8;
  delay();
  PORTB+=8;
  delay();
```

```
    PORTB-=8;
    delay();
    PORTB+=8;
    delay();
}

// amber and red on
void amber02(void)
{
    PORTB=20;
    delayMedium();
}

void main()
{
    bitclr(STATUS, RP0);  // Select Bank 0
    bitset(STATUS, RP0);  // Select Bank 1
    TRISB = 0x00;         // Set Port B pins to output mode
    state01();            // red on one side and green on the other
    blinkGreen01();       // green light 01 blinks
    amber01();             // still red on one side and amber on the ot
her
    state02();            // this state is like state 01 but inverted
    blinkGreen02();       // green light 02 blinks
    amber02();             // amber light on one side and red on the ot
her
}
```

When the simulation is running we can see the complete sequence as a representation of a working traffic light. First the cars in one street have the red light on while the cars on the other street can cross safely. After a given time (defined in the `delayLong()` function) the green light starts to blink before going into amber and finally to red. At this time the traffic light allows traffic on the other street while stopping the cars on the other street.

This example concludes the chapter on Multisim capabilities for simulating microcontrollers. We went through the steps on getting an appropriate environment for writing and compiling programs, both in assembly language and in C. Then we showed a set of applications that involved a program being compiled and transferred to the microcontroller and external peripherals being connected to it.

4.6 CONCLUSIONS

A simulator that handles a mixture of digital and electric circuits is a very powerful tool. A designer can benefit from such software the moment he is free to explore new variations on the design without the worries of time-consuming recalculations.

This chapter joins the analog/digital world with the world of microprocessors, the hearts of today's devices, gadgets and appliances. As we have shown, the inclusion of microcontrollers in other types of circuits is a seamless job in Multisim.

Through examples we have show the reader some of the possibilities that Multisim opens. The reader should feel free to try more complex programs and connections to more peripherals or external circuits if so he/she desires.

REFERENCES

[1] S. Oualline, *Practical C Programming*, O'Reilly, Sebastopol, CA, 1991.
    Cited on page(s) 98

CHAPTER 5

PCB Design With Ultiboard

INTRODUCTION

An electric circuit can be analyzed numerically, can be tested in a laboratory with real instruments, can be simulated with software tools like Multisim or mixing hardware and software tools like Elvis. But in order to make an electronic design become a physical product –commercial and long-lasting—a manufacturing process is needed to create a printed circuit board or PCB.

A PCB is a design (much like a drawing) that can be used to transfer or *etch* the form and geometry of the conductive tracks that connect the electrical components involved. These tracks are soldered to the connection pins of each component or connector so the designed circuit works as expected. Multisim has a tool named Ultiboard which translates schematic circuits from Multisim to a PCB.

5.1 BASICS OF PCB DESIGN

Designing PCBs involves the ability to visualize electrical circuits from a different point of view. That point of view is the way elements are connected in a breadboard. This type of visualization is very similar to what electrical engineering students experience when they learn how to translate the electrical representation of a circuit in a schematic drawing and the final connection made in the breadboard.

The design methodology proposed here is one among several possible ways of arriving to a final product. It is a road full of alternative exits. In this chapter we show two similar methods that can be adapted to the needs or preferences of each user.

5.1.1 LET THE COMPUTER DO THE ROUTING

The first method is conceptually simpler because it lets the computer do the tedious task of routing the tracks according to a schematic design. This allows users to concentrate in the placing of each component instead of planning each copper segment individually.

This option is very quick but it has a practical drawback: it almost always assumes that the design is laid out in a double-sided copper board and, in some cases, in boards with intermediate layers of copper. For commercial products, manufactured automatically, this does not represent any problem; in fact, most of the commercial designs are multi-layer so complex connections can be done in a limited space.

Once we have pondered the aspects of automatic routing it is now time to show the procedure. The steps are as follows:

- The user enters the design in a computer as a schematic drawing.

- This schematic design can be simulated and/or debugged with the techniques explained in this book and in the companion book Circuit Analysis with Multisim [1].

- When the circuit has been simulated—or even when it is being entered into the computer—it is necessary to verify that each component has an adequate *packaging* defined. (Packaging is a label that describes how the component is *physically*. Specifying this value Ultiboard knows the dimensions, the pin spacing and the geometry of each device).

- A *netlist* is exported from Multisim to Ultiboard. The netlist is a text file that contains the connections between components and the packaging of each one.

- The netlists are imported into Ultiboard and the placement of the components is made (automatically or user-defined). This step is harder than it seems on first inspection.

- Routing options are selected and the autorouting is started.

- Depending on the results the final tracks can be rearranged, modified or routed again.

- Finally, the final design is exported in a format known as *Gerber*. This is the file that a manufacturer needs when making large production runs of a design.

This is the outline of the steps needed for having a PCB design made with Ultiboard and the autorouting option. In Section 3 we see a practical example of this method.

5.1.2 MANUALLY ROUTING A DESIGN

This method is a little bit harder for a beginner. The learning curve is steep, but the results are more than adequate for making single-side prototypes.

This method uses Ultiboard like a footprint library for components. As mentioned above, footprints are the physical designs associated with each packaging. When using Ultiboard for manually routing a design, these footprints are laid out anywhere on the PCB and the user can decide the width of each track, the routing and the labels for each electrical connection.

The layout and the spacing between elements largely depend on the specific needs of a particular design. As a starting point, the recommended guidelines for making prototypes with PCBs can use the following values (in thousandths of inches or mils[1]):

- Hole diameter: 40 mils.

- Track width: from 10 to 60 mils.

- Circular pad width: from 80 to 120 mils.

- Width or height for rectangular or square pads: 80 to 120 mils.

These dimensions are well-suited for PCBs that are relatively easy to convert in prototypes and that can work with through-hole components, as their use is widespread in many applications.

The procedure is as follows:

- Start Ultiboard.

- Define the board size you plan to use.

- Insert into the design the footprint for each component.

- Components are laid out as the user needs (with practice is easy to spot problematic areas, like grouping pins for interconnections and/or divide the area in functional zones).

- When the components have been placed on the board, we need to connect each node. This step is similar to the first method when the computer is ready to start its autorouting process.

- Instead of instructing the computer to start routing, the user should start making the electrical connections drawing copper tracks between pads. A schematic diagram is certainly useful and the netlist file can be a guide for verifying which nets are still left unconnected.

- The connections should be made using the widest setting for the track width. When the track should go through a reduced space, it is perfectly acceptable to reduce the width only in that segment. Using a wide track width makes the design easier to transfer and etch when doing prototypes.

- When routing is finished, the design can be printed and transferred to a copper board using whichever method the user prefers (photosensitive emulsion, lithography, transfer sheets, Gerber or DXF files for automatic manufacturing, etc.)

Both methods have several common points. We discuss an example with notes indicating whether a step belongs to the automatic or to the manual method.

5.2 STEP-BY-STEP EXAMPLES

When a user needs to make a PCB, the pre-requisites are:

- A schematic diagram showing the electrical connections between elements.

- A list of the corresponding package and footprint for each device or element.

In this section we show two examples, the first one uses the automatic method of letting the computer do the routing itself. The second example cannot be reliably routed by the computer when a single-layer board is selected; this forces us to do manual routing of the tracks.

Example 5.1 Designing the PCB for an inverting amplifier.
A very popular IC in the lab is the operational amplifier TL081 (and related members, like TL082 and TL084). This IC is an operational amplifier equivalent to the venerable 741 and its pins are directly compatible with it. In this example we design the PCB for an inverting amplifier with a gain = 100.

5.2.1 SELECTING THE OPERATIONAL AMPLIFIER

The first step is selecting the right component. From the datasheet we can see that the packaging for the op-amp we have here is DIP or PDIP. These are the initials for Dual In-line Package or Plastic Dual In-line Package. These packages are some of the most common types available in integrated circuits. Manufacturers prefer surface-mount components, but for a prototype they are much harder to work with, so this DIP is perfect with its pins that can be soldered from the underside of the PCB.

Once we have gathered the packaging information we can start to draw the schematic for the inverting amplifier using Multisim. In the Place → Component menu shown in Figure 5.1 we select the family for analog components and in the OPAMP section we find the model for TL081. Notice the models available in the Model manuf./ID window. We need to select the one with the PDIP-8 designation in the Footprint manuf./Type window.

5.2.2 SELECTING RESISTORS

The next step is the selection of the resistors. Thus, we select a resistor with footprint RES1300-700x250 as can be seen in Figure 5.2. This is a rather big footprint but very didactic and easy to handle.

5.2.3 SELECTING HEADERS

Since this is a circuit that is going to be used in the physical world, we must not forget to include headers. Headers are connectors used to hook external components to the PCB. For this example we use two HDR1X3 connectors that can be found under Place → Component → Basic → Connectors. This opens Figure 5.3. The footprint is marked as generic and later we see in Ultiboard how to handle these connectors. In the first one we wire Vin, Vout and GND. The second connector has the power lines coming from an external power supply: +VCC, -VCC and GND.

The schematic drawing for the final circuit is shown on Figure 5.4. We save the circuit file with the name Example01_Inverting_Amplifier.ms11.

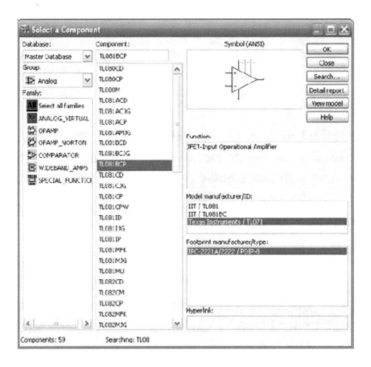

Figure 5.1: Selecting the operational amplifier.

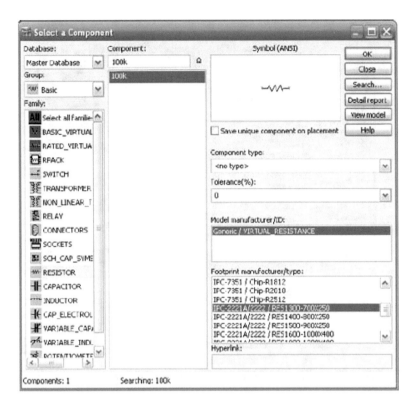

Figure 5.2: Selecting the resistors.

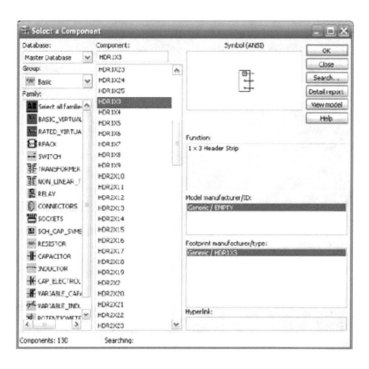

Figure 5.3: Selecting the headers.

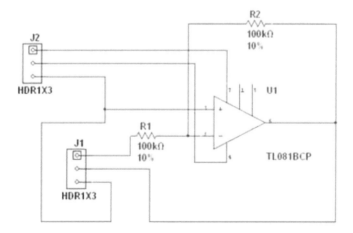

Figure 5.4: Schematic drawing for the inverting amplifier with connection headers.

5.2.4 TRANSFERRING THE DESIGN TO ULTIBOARD

This step shows us how to transfer the design to Ultiboard. Depending upon the installed version available we should choose to send the design to Ultiboard version 11 or previous versions. This is done in the menu Transfer → Transfer to Ultiboard → Transfer to Ultiboard 11.

As a first requisite, Ultiboard asks for a file name. In this case, the file name is chosen as Example01_Inverting_Amplifier.ewnet. Note that Multisim creates a file with the extension ewnet. This is the file that Ultiboard uses to create the PCB. This launches the program and we see a window like the one shown on Figure 5.5. This is the import netlist which describes how the elements are connected in the circuit.

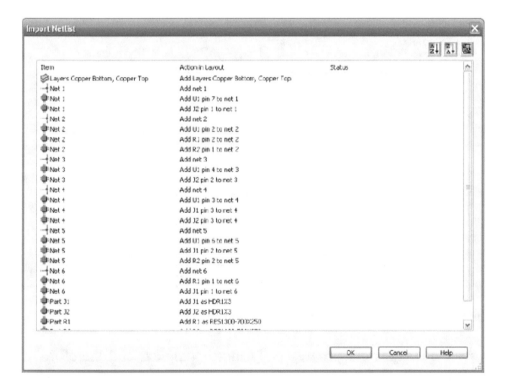

Figure 5.5: Import netlist.

As we can see in the main Ultiboard window, the tracks are 10 mils in width. This is the default width. We can change the units but the recommended units are mils. This is due to the fact that the vast majority of components are manufactured using mils as a reference, including the pin spacing of holes in a breadboard, so it is a safe procedure to work using the same units as the component packages we are using. From the menu Options → PCB Properties, the window for the PCB properties opens and in the Design rules tab we see the track width set at 10 mil. The recommended value when working with simple designs is 25 mils.

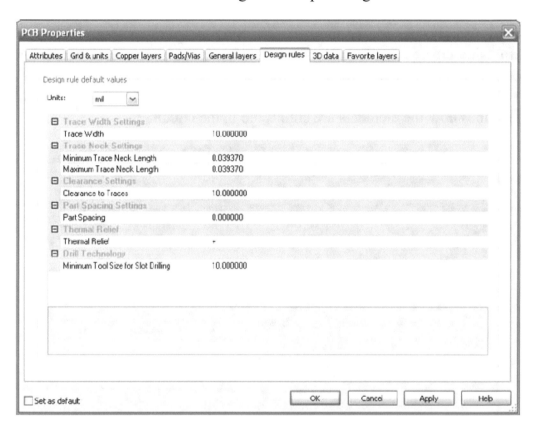

Figure 5.6: PCB properties. The Design rules tab is selected.

As the designs get progressively more complex, and therefore, more dense, the track width can be reduced to 10 mils or even less. The last parameter, *clearances to trace* is the minimum distance between contiguous tracks. The value of 10 mils is appropriate for our needs. Once we have imported the netlist we get the main window for Ultiboard, with the board outline and the imported components placed *outside* of the board (Figure 5.7).

We can zoom in and out with the mouse scroll wheel so we get a better view of the components and their connections. The yellow lines in Figure

5.8 are the imported nets; these allow the autorouter to know which component is connected to another one.

Here we can see the labels for each component; for example, J1 and J2 are two headers, U1 is the DIP8 integrated circuit and R1 and R2 are the resistors. The yellow lines, as we explained above, are the electric connections between elements; these come from the netlists and due to their entangled look they are sometimes called *ratsnest*.

In the View → Toolbars → Select menu we can activate the selection toolbar shown in Figure 5.9.

The first button enables the option for selecting components. When enabled we can drag components to any convenient location before routing. We can also enable the autoplacement function in the *Autoroute* toolbar but we must take into account that the results are not always what we are expecting. In this example we place the components as shown in Figure 5.10.

Figure 5.7: Initial placement of footprints in Ultiboard.

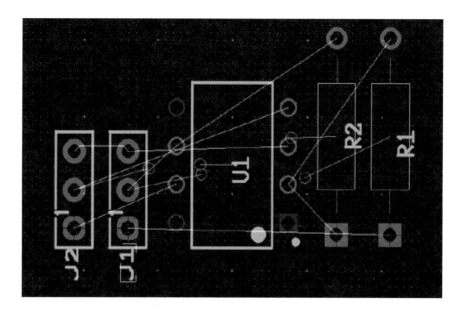

Figure 5.8: Zooming in to see a detailed view of the components.

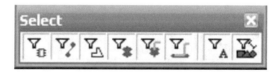

Figure 5.9: Selection toolbar.

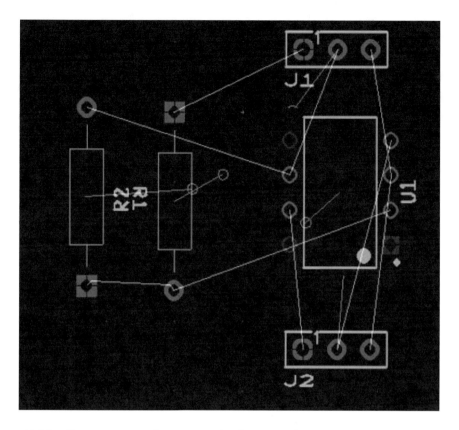

Figure 5.10: Component placement before routing.

Now that the components are in place we can choose to use a manual placement of the tracks or the autorouting utility. First we see what happens with autorouting. The autorouter is invoked in the Autoroute → Start/Resume Autorouter menu. The results are similar to those shown in Figure 5.11.

As we can see, the results are very good. We could check to see that each connection is done properly to the corresponding pin in the IC.

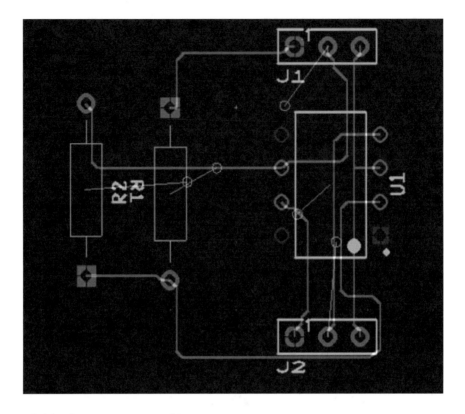

Figure 5.11: Autorouter results.

For a simple design the autorouter works well. For complex designs there is a high probability of tracks not being routed. In that case, we must manually draw the tracks.

Example 5.2   Signal conditioning for a digital thermometer based on LM35.

We are going to design a PCB for a digital thermometer based on the LM35 integrated circuit. This circuit has a TO-92 packaging typical of small-signal transistors. Its three terminals are VCC, GND and Vout. In its basic mode of operation we get 10 mV for each Celsius degree and it is capable of handling a temperature range of 0 to 100 °C.

We start with this range in mind and assume that we need to feed this voltage to another circuit that requires a range of 0 to 5 VDC, where 0 V is equivalent to 0 °C and 5V is equivalent to 100°C. This is a pretty common scenario in real life, where an analog signal is the source of a digital circuit that operates with TTL voltage levels.

The required signal conditioning involves a voltage gain of 5. We can then use an operational amplifier with a DIP8 packaging, for example a TL072. This circuit has the advantage that internally there are 2 independent op-amps and then we can use an inverting amplifier with a gain of 5 with a variable resistor for calibration and an inverter amp with unity gain for a positive output. The complete circuit can be seen in Figure 5.12.

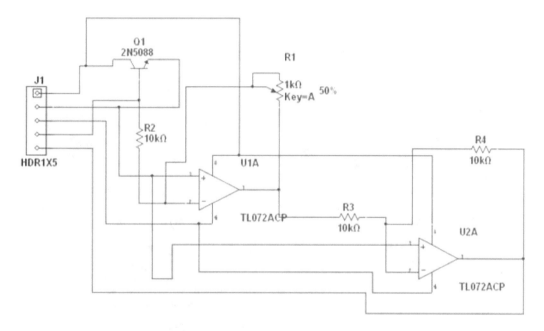

Figure 5.12: Signal conditioning for an LM35 sensor.

Notice how we are using a 2N5088 BJT transistor instead of the LM35 sensor. This is done because they both share the same TO-92 packaging and there is no LM35 in the parts bin.

Again, we are going to transfer the design from Multisim to Ultiboard (from the Transfer → Transfer to Ultiboard → Transfer to Ultiboard 11 menu). Once we are running Ultiboard we are going to accept the default configuration and we have the components placed outside the board outline of our PCB.

With this default configuration we can make a test to see the results for a fully automated design. First, we need to autoplace the components (Autoroute → Start Autoplacement menu) and with that result we are going to try the autorouting feature (Autoroute → Start / Resume Autorouter). In our particular case we obtained 16 out of 17 tracks routed. This is a 94% of the whole design. In this basic design that means that we have to manually route one track or maybe use a wire jumper to make the connection. In a more complex design this could probably mean that we have to manually route dozens of tracks.

In this case, we can place a pair of *vias* on the board. Vias are connectors for going from one copper layer to another. Here we are going to use them in order to go from the copper top layer to copper bottom layer.

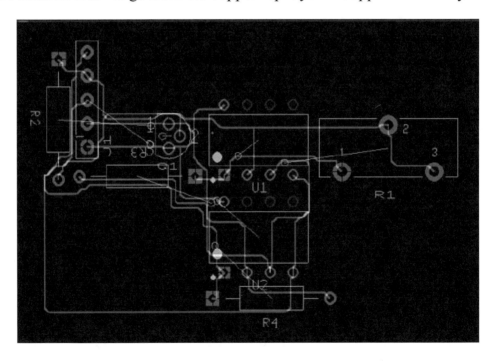

Figure 5.13: PCB connection using vias.

Here it is evident that there is a need for a mixture of routing and placement techniques. Each designer is going to develop his/her own set of techniques for the tracks that could not been routed by the software (see Figure 5.13).

In these examples we can see the design possibilities that Ultiboard offers. The next step in the PCB fabrication process is the electrical rules check. This is a check-up that can be run when a design is apparently finished; we can specify some criteria like distance between adjacent tracks and minimal width of tracks. This is useful when a design involves the use of RF (radio frequency) or high frequency signals.

In Ultiboard we can activate the Electric Rules Check by going to the menu Design → DRC and Netlists Check. If we check the design we have so far, we are hopefully not going to find any errors, but in more complex circuits it is always a good practice to periodically check for errors so we can fix them as they appear. This is going to save us a great deal of time in the long run.

If we wish to check the details or change the preferences for the electric rules, the menu we use is found under Options → PCB Properties → Design Rules.

A pair of interesting options that the Educational version of Ultiboard has are the PCB Transmission Line Calculator shown in Figure 5.14, and the PCB Differential Impedance Calculator shown in Figure 5.15. They are available in the menu Tools → PCB Transmission Line Calculator and Tools Tools → PCB Differential Impedance Calculator. Both are used for doing the calculations related to the electromagnetic values belonging to the material and dimensions used in our designs.

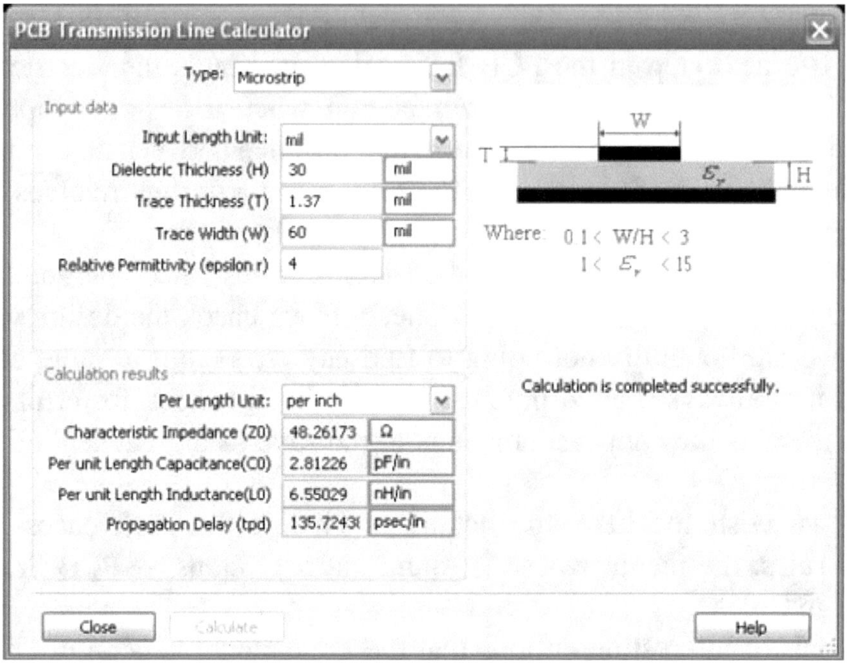

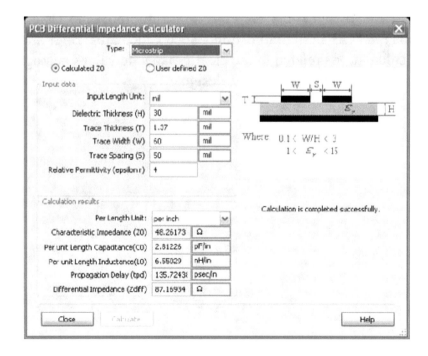

Figure 5.14: PCB Transmission Line Calculator.

Figure 5.15: PCB Differential Impedance Calculator.

5.3 FABRICATION

The final step in the process of making a PCB can be done using a number of very different techniques. For students making circuit prototypes it is very common to use techniques such as transferring with an iron, silkscreen or directly drawing the designs on the copper layer with a permanent marker.

For more complex designs, or designs involving two or more copper layers a quick alternative, but expensive, is the use of CNC (computer numerical controlled). These machines can make circuits with a high level of precision and practically with absolute repeatability.

Ultiboard allows us to take either route. For the simplest case the design can be printed on paper or in a number of transfer sheets available commercially. This is done under the menu File → Print and there we can select the desired size (a size of 100% is highly recommended when not using optical techniques like silkscreen), the printer, and the copper layers that we are going to print.

The procedure for the second case is different. We need to select the option File → Export and there we see a menu with different options for exporting files, among them SVG (Scalable Vector Graphics), Gerber and DXF (see Figure 5.16).

Figure 5.16: Export options.

If we select SVG, some design programs can handle very well the vector images obtained. As a suggestion you can use a free program called Inkscape that natively works with SVG files.

For Gerber files we need to specify which layers are going to be exported and after that define the *aperture* values. This is a value used by CNC machines for calculating the outline that should be routed when making the PCB. DXF export is similar with the difference that the user does not have to specify the apertures. DXF is a file type handled by CAD software.

5.4 CONCLUSIONS

In this chapter we have presented a few techniques for making PCBs using Multisim as a base for schematic capture and Ultiboard for designing the copper tracks.

We have also shown some variations on the basic techniques and the results that can be obtained. Finally, we listed the export options available so the user can choose the most appropriate for his/her needs.

REFERENCES

[1] D. Báez-López, F. E. Guerrero-Castro, Circuit Analysis with Multisim, Synthesis Lectures on Digital Circuits and Systems, Morgan and Claypool Publishers, Vol. 6, No. 3, 2011. Cited on page(s) 116

---

[1] A mil is a thousandth of an inch.

Authors' Biographies

DAVID BÁEZ-LÓPEZ

David Báez-López was born in Puebla, México. He attended the Universidad Autónoma de Puebla where he obtained a B.S. in Physics. He then obtained M.S. and Ph.D. degrees in Electrical Engineering from the University of Arizona. He has authored more than 80 research papers and 5 books. He has been a professor of Electronics at Universidad de las Américas-Puebla, UDLAP, in Cholula since 1985 where he was Head of the Department from 1988 to 1996, at the National Institute for Astrophysics, Optics, and Electronics, from 1979 to 1985, where he was also head of the Department of Electronics from 1983 to 1985, and he has been a visiting professor and researcher at Texas Tech University at Lubbock, TX, and at Ryerson University, Toronto, Canada. He is founder of the International Conference on Electronic Engineering CONIELECOMP, held every other year at UDLAP.

FÉLIX GUERRERO-CASTRO

Félix E. Guerrero-Castro is an electronics engineer working at Hackerspace (`http://www.hackerspacecholula.org`) in Cholula, Puebla, México. Among his daily activities are hardware development, consulting, web programming, and teaching kids how to program Arduino microcontrollers.

After receiving his Master's Degree in Electronics in 2004, he spent six year as a lecturer teaching electronics at the Electronics Lab at Universidad de las Américas-Puebla UDLAP. He holds a Bachelor's degree in Electronics and Communication Engineering from the Instituto Tecnológico y de Estudios Superiores de Monterrey (ITESM) 1999. He has published a number of research papers in power electronics, digital sound modelling, and biomedics. He also plays guitar and records local bands in his own studio. He enjoys traveling and taking landscape photographs.

OFELIA CERVANTES-VILLAGÓMEZ

Dr. Ofelia Cervantes-Villagómez holds a B.S. in Computer Systems Engineering from Universidad de las Américas, Puebla, (May 1981), a Master of Sciences in Computer Systems from l'École *Supérieure en Informatique* in Grenoble, *France*, (June 1984), and a Ph. D. degree from l 'Institut National Polytechnique de Grenoble, France, (January 1988). Dr. Cervantes-Villagómez is co-founding member of the *Laboratorio Nacional de Informática Avanzada*, A.C. (National Laboratory for Advanced Informatics, A.C.) established in Jalapa. Ver. She was a member of the Directive Board for an eight-year period. Starting in the year 2000, she was designated by CONACYT to promote cooperation between Mexico and France and Germany to pursue research in Information Technologies. She participated also in the creation of the Mexican-European Informatics Laboratory, which has as collaborating countries Mexico, Spain, France, and Germany. Dr. Cervantes has experience as a professor, researcher and has consulted in Data Base Modeling, Distributed Systems, Artificial Intelligence, Automatic Speech Processing and Business Intelligence. She has taught a number of short courses and has many publications in these topics. She was a visiting research fellow in IMAG, Grenoble France (1995) and Université du Québec à Trois Rivières (2005). She has also been thesis director to several undergraduate and graduate students at the M.S. and Ph.D. levels. Dr. Cervantes-Villagómez was president of the Mexican Artificial Intelligence Society, President of the Mexican Association for International Education, as well as founder and president of the first Chapter of the International Association of International Educators Phi-Beta Delta. Dr. Cervantes-Villagómez is still active in teaching and research at UDLA. She is also the Honorary French Consul in Puebla since 2003.

Printed in the United States
by Baker & Taylor Publisher Services